Jeff Ritter, PGA

"Michael Hebron is doing incredible work. Much needed for the future and well being of the game. Coaching is about "people" and his mission is needed now more than ever!"

"Information without application is useless. Michael Hebron's learning-based research provides coaches with the tools most critical to stimulating positive action towards a winning result. His impact will prove to be among the most powerful and positive influences on those who teach and play the game."

•••

AJ Sikula

"I experienced an incredible day observing the living legend, Michael Hebron. Brilliant eye for the golf swing, but HOW he presents his information to students is light years ahead of others and is the future of how to teach, applicable to all industries, not just golf. Great stuff!"

JOINING THE ART OF TEACHING WITH
THE SCIENCE OF LEARNING

LEARNING with the BRAIN IN MIND:

MINDSETS BEFORE SKILLSETS

BY MICHAEL HEBRON
WITH STEPHEN YAZULLA, Ph.D.

"HOW WE LEARN IS DIFFERENT FROM HOW WE THINK WE LEARN"
DR. ROBERT BJORK, UCLA LEARNING & FORGETTING LAB

Learning with the Brain in Mind, Mind Sets Before Skill Sets

First Edition–published in 2017 by Learning Golf, Inc.

495 Landing Avenue, Smithtown, New York, 11787

(http://www.michaelhebron.com) and
(http://www.nlglive.com) aka Neuron Learning for Golf Live

ISBN 978-1-937069-08-7

Editorial Assistance by Nannette Poillon McCoy

Cover Design by Nicole Hebron

Library of Congress Cataloging-In-Publishing Data

Hebron, Michael

Learning with the Brain in Mind, Mind Sets before Skill Sets/Michael Hebron,
1st Edition

Includes bibliography references

Library of Congress Control Number: 2017934274

First Edition–published in 2017 by Learning Golf, Inc.
495 Landing Avenue, Smithtown, New York, 11787
(http://www.michaelhebron.com) and
(http://www.nlglive.com) aka Neuron Learning for Golf Live

**THE PURPOSE OF
APPROACHES TO LEARNING
IS
TO SERVE THE NEEDS OF THE
BRAIN,
WHICH IN TURN SERVES
THE NEEDS OF THE
APPROACH.**

Contents

Preface .... *a*
Introduction .... *i*
Opening Insights *1*
Myths and Facts *21*
Views on Learning and Teaching *35*
Early Educational Research *43*
The Nature of Learning *51*
Studies, Strategies, and Research *73*
Learn, Develop, Grow *vs.* Teaching-Fixing *103*
Approaches To Learning Are Different Than Approaches To Teaching *113*
Emotions *135*
Basic Look at Memory *153*
Closing Thoughts *161*
My Journey *171*
About Michael Hebron *175*
Acknowledgements *177*
Bibliography *179*

Preface

By Stephen Yazulla, Ph.D.

One might ask why an internationally renowned teacher compiled studies about improving acts of learning, teaching, and the brain's connection to both.

Michael Hebron questioned; Why were some individuals experiencing meaningful learning while others were not? What should we know about the nature of learning? How should students be evaluated? Is there a need to rethink the relationship between learning and teaching? Hebron also questioned what his own blind spots may be when it came to the topic of learning.

In general, current methods of teaching, regardless of the topic or setting, emphasize content, memorization, drills, practice, and test taking. Some approaches to teaching tend to look for what is broken and attempt to fix it. An alternative is suggested in *Learning with the Brain in Mind*; provide a safe, playful, less judgmental environment in which self-discovery, experimentation, and adaptation are encouraged. An alternative that is based on the brain's connection to the nature of learning.

In Hebron's view, when we engage anyone's curiosity you will have their attention and motivate them. St. Augustine, referring to his resistance to learning Greek, stated "This experience sufficiently illuminates the truth that free curiosity has greater power to stimulate learning than rigorous coercion." Confessions, p 17

What is obvious to us now, but not so even 150 years ago, is that with all our behavior, including learning, inputs from the senses, output to the muscles and the coordination of this activity takes place in the brain first.

The efficacy with which this coordination takes place depends on numerous factors in the environment, including: potential effects (payoff), probability of success, social setting, sense of danger, whether we are thirsty, cold, hot, hungry or fatigued, all of which have profound effects on the brain and how well one learns and performs.

Each of the above factors influences the internal organization of the brain, some of which enhance learning, while others interfere with learning. Conditions, in which learning is enhanced, are referred to as "Brain-Compatible." In this view, the biology of the brain and what we call "Mind" are intimately connected.

The field of Brain Based Learning has grown significantly with the introduction of new technology allowing us to better understand how the brain functions, the effects of varying circumstances on learning, and the brain's connection to those processes of change.

"A major effort in schools and other training institutions is directed at improving the conditions in which learning occurs in order to increase the efficiency of learning. The obvious question then is how to determine if learning has taken place, and if so, how much and how long will it be retained. Learning cannot be measured directly. It is inferred by observing PERFORMANCE. Performance engages the motor system involved in speech, running, hand movements, etc. This means that regardless of how new information has been encoded and stored in the brain ciruits, performance is limited by the motor circuits used to express learning.

For example, if you were to memorize a poem, the only way to determined if it was learned is to openly recite or write it from memory. Thus, even if you could silently recite it, your external expression of it is limited by your ability to speak or write. Stage fright or butterflies would adversely affect how you recited the poem, and I would conclude that you had not learned it very well. Performance is how learning is assessed.

This book is arranged by first presenting some of the terminology used in this book to enable all readers to have a common vocabulary and appreciation of the interaction of the nervous system to conditions that affect learning. Terminology is followed by the scientific sections of *Learning with the Brain in Mind.* These are a series of chapters that provide insights into how Brain-Compatible Learning can be accomplished.

Introduction

Learning with the Brain in Mind takes into consideration that the brain is our gateway to learning, and what we do mentally and physically is organized in the brain first. What has been compiled here is influenced by many sources including insights that were obtained from speakers and attendees at conferences I attended. It is also influenced by my many notes taken on the topic of learning from journals, research papers; at workshops and conversations I had with leaders in the field of learning, and by years of personal experience as an instructor. My curiosity about the nature of learning drove me to discover insights that would dramatically change how I was interacting with students. A review of the terminology used here follows.

TERMINOLOGY

Meaningful Learning

Brain-Compatible Learning

Metacognition

Motivation

Safe™, Smart™, Playful™

Relate-Remember-Repeat

Coaching - Teaching

The Brain

In this book, any provider of information, regardless of the circumstances, is referred to as an "instructor" or "teacher." Similarly, any receiver of information will be referred to as a "student" or "learner."

MEANINGFUL LEARNING (ML)

Meaningful Learning refers to the kind of learning that has been encoded long term. It is information or skills learned in one environment that can also be used in a variety of dissimilar situations. Meaningful Learning is based mostly on what we knew before we experienced new learning, which is why metaphors and storytelling in learning environments are such powerful tools.

"This ability is referred to as 'generalization.' It means to transfer knowledge from one situation to another." (Professor Stephen Yazulla)

When Meaningful Learning is brain-compatible it is:

Joyful and fulfilling	Open-ended and direct
Personal and satisfying	Flexible and portable
Adventurous and safe	Inventive and personal
Interesting and challenging	Internal and democratic
Autobiographical and stimulating	Plain spoken and tinkering
Unconscious and lasting	Wordless and seeing
Indirect and spontaneous	Understood and felt

When information-delivery systems accomplish these views of learning, human flexibility and fluid intelligence are leveraged, guiding students in the direction of seven C's: comparing, combining, connecting, calculating, clarifying, creating, and changing their views of information.

Approaches to learning that support meaningful long-term improvement are less about avoiding, or reacting to, or overcoming unworkable outcomes and more about being active and proactive about learning what needs to be accomplished.

ML environments can develop the ability to hear what is not being said, to see what is not being shown, to read between the lines, and to have answers for questions yet to be asked.

> ***Often, in an ever-changing environment, ML is being used unconsciously. New York University Professor Joseph E. LeDoux, Ph.D. states, "Conscious recognition of unconscious learning is meaningful learning."***

Innovation is the engine of ML. If approaches to learning create environments that fail to promote innovation on the part of a student, it would be akin to making a car lighter by removing its engine.

ML often does not appear to be a composite result, but an outcome brought on by spontaneous interaction developed from a variety of different skills in ever-changing conditions. ML is a product of the brain's ability to filter new information through the lens of prior knowledge and experiences consisting, of both workable and unworkable outcomes. This process gives meaning to new information.

The view here is that ML manages and increases existing levels of talent. Insights about talent management in The Peoria Journal Star (2/21/2014) highlighted the importance of cross training by asking three questions:

> ***How can talent best be identified?***
>
> ***How can we anticipate future talent needs?***
>
> ***How can talent be best molded for future use?***
>
> ***One Answer: Use the kind of ML approaches that expose talent to multiple diverse environments.***

Curiosity may be the most useful way of looking at motivation and the nature of ML. Keep students curious, and they stay motivated. Learning-developing approaches are interested more in the way people are thinking creatively than what they have done when following the kind of directions found in teaching-fixing to get something right approaches.

When is information not supporting ML? Perhaps a common response would be "when information is false." However, many false ideas, poor answers, and inconsistencies have been the starting point for ground-breaking discoveries. For example, the glue on Post-Its® was an accidental discovery in 1968 by Dr. Silver Spencer of 3M®, who was attempting to develop a super-strong adhesive. Instead, Dr. Spencer produced a low adhesive, the tacky glue that eventually was patented and found near universal use on the reusable Post-Its®.

> Since expectations influence ML and influence gaining a good education, these are questions for both instructors and students to consider.
>
> What expectations do individuals have about their learning environment?
>
> What expectations do individuals have when learning?
>
> What knowledge and skills are needed to be an efficient learner?

The components of meaningful learning include emotional, social, cognitive, physical, and reflective components. All five components function simultaneously. No component can be completely turned off. Each component's action affects all the others as part of the whole. Every question or problem will touch all components.

Emotional = Safe, personal insights and interests.

Social = Collaborate, self-evaluation, authentic, personal decision making.

Cognitive = Reasoning skills based on one's experiences.

Physical = Active bodies and minds learning, not passively following.

Reflective = Makes all other components efficient with non-conscious records of experiences.

To be encoded, long-term information must connect with all these components seamlessly. For example, information must touch emotions and create interest, be challenging, be personally relevant, connect with past knowledge, and be independently complete. Emotions must be safe to support personal interests that lead to fun, and challenges that create growth.

MEANINGFUL LEARNING (ML) is not fully supported when new information:

Does not promote self-discovery—is not developmentally appropriate

Is not transferable to multiple contexts

Is outcome, not process orientated

Is not part of a story or metaphor—is presented only one way

Has too many details

Causes frustration and intimidation—creates more stress than curiosity

Has no meaning to students—cannot be related to an experience

Makes no sense to students—is out of context

Does not support self-confidence—does not reduce self-doubt

It appears that meaningful learning comes from the best ingredients on earth, curiosity, imagination, and improvisational skills and not from criticisms or corrections, or trying to get it right.

MEANINGFUL approaches to learning move beyond just knowing the subject content.

STUDENT-CENTERED/BRAIN-COMPATIBLE LEARNING

Brain-compatible Approaches (BCAs) impart a methodology of learning to students. They help students sharpen their intellectual capacities. They support curiosity, creativity, inventing, and execution, not just ideas. They put to use a mind-body connection that is influenced by emotions. They are student-centered and brain-compatible.

BCAs to learning are geared for personal development through stages, not for perfection. They look for participation, not perfection. Learning, development, and growth through stages support encoding long-term learning.

Student-centered instruction [SCI] is an instructional approach by which students influence the content, activities, materials, and pace of learning. This learning model places the student (learner) in the center of the learning process.

Student-centered learning is also known as learner-centered education. Student-centered learning broadly encompasses methods of teaching that shift the focus of instruction from the teacher to the student.

Suggestion! *A student-centered/brain-compatible approach to learning embraces growth with an open mind and experiences progress from both workable and unintended outcomes.*

BCAs to learning support the development of individuals who become confident, eager, self-motivated learners; individuals who are independent thinkers, curious, and self-reliant; learners who gain self-knowledge, while fulfilling their potential with open minds.

Student-centered/brain-compatible information-delivery systems provide a message of inclusion. They join prior knowledge with new information and join non-judgmental words with unworkable outcomes. Too often this is not the case.

Traditional teaching approaches tend to pay more attention to information, instead of how we are going to use the information in a context of changing circumstances. Keep in mind that acquiring, categorizing, applying information, and reacting to outcomes is mostly in the hands of our non-conscious mind.

Student-centered/brain-compatible information-delivery systems are caretakers of learning skills first, and instructors second.

BCAs to learning do not condemn instructors to be dictators, by which acts of teaching and fixing can suppress a student's will to be curious and confident. When student frustration and intimidation are removed from approaches to learning, a journey of meaningful learning follows.

When information-delivery systems are student-centered/brain-compatible, human flexibility and fluid intelligence are leveraged, guiding students in the direction of changing their views of information. Student-centered/brain-compatible learning environments turn learning into pleasure, and pleasure into learning, with an emotionally safe approach.

BCAs to learning are not a fad or just another educational experiment. They represent meaningful changes from traditional approaches to learning based on the brain's connection to learning.

A student-centered/brain-compatible statement by an instructor, "Hi, I am here to help you invent your math, reading, writing, thinking, decision making, or sports skills."

"Though we often say that we learn in school, learning takes place in the brain," said Kenneth Wesson whose work in education, learning and teaching spans four decades.

New information is evaluated by the brain before it becomes new learning or it is set aside as not useful. This process is mostly an internal non-conscious process based on experience(s) that asks "does this information make sense and have meaning?" These are two requirements of the nature of learning anything.

> ***Let's stick with the message here—a brain's connection to learning provides a winning formula!***

When it comes to BCAs, cognitive scientists suggest that emotions influence learning more than any other element found in the nature of learning.

In BCA learning environments, unworkable outcomes are valuable and necessary feedback for future reference and not in need of fixing. Because learning is about change, some incorrectly feel that learning is about fixing. Fixing is not learning.

Efficient approaches to learning are more about protecting and developing natural skills and talent than the negative steps of fixing anything.

Change can be introduced by learning what to do, or what is required while avoiding any thought about fixing a poor habit or outcome. BCAs to learning join the art of teaching with the science of learning.

In BCA learning environments, unintended outcomes are more useful for improving and developing skills than workable outcomes. Having students make and then repeat errors during lessons is a useful tool for improving and learning how to produce a different outcome. The nature of learning uses the sensory feedback between workable and unworkable outcomes to facilitate awareness of the differences by positive reinforcement.

BCA learning environments are more about what we can do with what has been learned, than knowing detailed information about a topic.

BCA learning is about what can be accomplished, not about overcoming some uncertainties.

In BCA learning environments, playfulness and imagination support a journey of learning that students would otherwise not have available to them. Meaningful learning is an adventure on the way to things that students may not have expected whether they be true or untrue.

BCA learning approaches are more about showing appreciation for growth, large or small, and less about appraisals of outcomes. These kinds of learning, change, and growth found in these approaches to learning are aimed at the future use of new information and insights beyond classrooms, business-training programs, and sports practice fields.

In BCA learning environments, instructors do not impose their expertise on students, they create a collaborative journey of growth and development. Educators, coaches, employers, parents, and instructors deliver information with excitement, compassion, and curiosity in student-centered/brain-compatible learning environments.

These BCA environments acknowledge the capacity for students to display their intelligence while they are still developing. BCAs can lower or remove frustration and intimidation while on a playful path to progress.

METACOGNITION

In the 1970s, John Flavell coined the term "metacognition," referring to our ability to evaluate our own thinking. This knack for reflecting on our thoughts is often viewed as the hallmark of the human mind. Ultimately, metacognition serves as a foundation for learning and success. (Scientific American Mind, September/October 2014)

Metacognition is a process that is important for experiencing meaningful learning and defined as "cognition about cognition" or "self-knowing about knowing."

The American Society of Training-Developing points out that metacognition is a set of high-level brain controls that guide each individual's information processing. Metacognition plays a role in the process whenever individuals set out to learn and problem solve. People use these skills from an early age and develop them further each time they are utilized. Metacognition is a skill set that includes five factors that separate successful puzzle solvers from unsuccessful learners.

Planning: the ability to organize time and resources effectively

Selecting: separating critical elements from non-important information

Connecting: linking new content to existing knowledge

Expanding: bringing new knowledge to a situation and adjusting prior knowledge accordingly

Monitoring: continually verifying understanding and application of knowledge

Note: **Fixing is not learning or investigating. Fixing acts of poor outcomes is not using the value of inconsistency and unconscious influences of metacognition.**

An example of our metacognitive capacity to regulate one's cognition to maximize the potential to think, learn, evaluate, and remember is a recognition of having more trouble learning "A" than "B." Another would be the thought to double check "C" before accepting it as fact. Human beings do not come into the world with fully developed metacognitive skills. We come into the world with the capacity to non-consciously learn these skills.

Any educator, guide, or coach who uses an approach that improves acts of learning, teaching, and performing, has found a way to enhance metacognitive skills. Coaches, educators, guides, employers, parents, and puzzle solvers need to develop metacognitive skills.

The term refers to the conscious and non-conscious mental activities involved in mental activities. It refers to mostly non-conscious, internal thinking involved in actively influencing learning in new situations. Examples include:

Organizing an approach to learning

Monitoring comprehension of the activity

Evaluating progress

Distraction awareness (the ability to become aware of distracting stimuli both internal and external)

Metacognition takes on many forms including insights about when and where to use a particular strategy for improving, seeing options and problem-solving. Efforts to enhance metacognition are arrived at by developing a learner's independence and self-regulation skills. This is classified into three components:

Knowledge: What individuals know about themselves

Regulation: Evaluating learning experiences through activities that help individuals influence their own learning

Experiences: Current, on-going cognitive experiences

Individuals with metacognitive skills are self-regulated learners who can modify learning strategies and skills. They are aware of blocks to learning as early as possible and change tools and approaches to ensure goal attainment. When tools and approaches are non-specific, general, generic, and not content dependent, they are likely to be useful in different types of learning situations.

Meaningful learning is often more about subtracting or changing poor insights, than adding information or fixing. Acts of enjoyable learning use hidden skills of metacognition that value subtraction of both information and physical actions. This process also develops personal knowledge about any topic.

For example, driving to a destination in a city for the first time without a GPS system can provide the advantage of getting a little lost and seeing parts of the area that the GPS would have avoided. When walking around this city later in the day, you would have insights and information for finding your way around the area that a GPS system would not have exposed you to.

Teaching-fixing to get it right approaches are like GPS systems; self-learning is being left out. On the other hand, learn-and-develop approaches know the value of getting a little lost and how this can develop future non-conscious reference points that support meaningful learning through metacognition.

MOTIVATION

In July 2014, Yale School of Management Professor Amy Wrzesniewski and Swarthmore College Professor Barry Schwartz published an op-ed, The Secret of Effective Motivation, in both the Proceedings of the National Academy of Science and the New York Times' Sunday Review (Gray Matter section) Wrzesniewski /Schwartz article pointed out two types of motivation; "internal" and "instrumental."

Conducting research for the joy of uncovering facts is an act motivated by internal motivation.

Conducting research to become well known or wealthy is an act motivated by "instrumental" motivation.

Some educators believe mixing these motives would be a good thing, but it is counterproductive per Wrzesniewski's and Schwartz's research. They reference studies that suggest efforts should be made to design and structure learning activities so that instrumental consequences do not become motives. Helping people see the meaning and the impact of their efforts, rather than gaining external rewards, may be the best way to improve the quality of their work and their ability to succeed.

Their studies show that when students are not that interested in learning, external incentives may prompt participation, but can result in less educated students when it comes to long-term retention. Trying to make an activity more attractive by emphasizing both internal and instrumental motives is understandable, but studies show this can have the unintended effect of weakening what is essential to success; internal motives.

Over 11,000 cadets in nine entering classes to United States Military Academy at West Point were questioned about their motives for entering the academy. Cadets with strong internal (for the joy of learning) and strong instrumental motives (for external acknowledgment) performed worse on every measure than cadets with strong internal motives, but weak instrumental ones. The cadets with strong instrumental motives were less likely to graduate, less effective as military officers, and less committed to staying in the military.

This paper points out that there is a temptation among educators and instructors to use what instrumental motivation tools are available to improve performance. While this strategy may create participation, it can detract from fully educating students. All educators should do their best to keep student curious and motivated.

Kathleen Cushman's *EIGHT CONDITIONS THAT IMPROVE MOTIVATION*

V x E = M (Value times Expectancy equals Motivation) If students value an activity and expect to be successful, they will be motivated. Kathleen Cushman's article, "What Kids Can Do," (published in Kappan Education, 2014), shares her equation on motivation. After interviewing hundreds of adolescents on the conditions that increase school motivation, she distilled the following:

Condition #1: We feel okay. "The stresses that students experience, at school or outside of it, take biological priority in the brain over learning," says Cushman. "Learning is very difficult when elemental sensations—fear, shame, hunger, exhaustion, loss, even distraction—stand in the way." Students need to feel safe and cared for. The school and classroom culture need to be supportive. It is important that teachers focus on inquiry rather than having students competing for the right answer.

Condition #2: It matters. Cushman states, "Even when the subject does not appeal, students are more willing to engage if it presents an intriguing puzzle or an issue of fairness." For example, a New York City student was unenthusiastic about a science unit on fueling the car of the future, but the student lit up when the teacher played an introductory video showing that oil is a finite resource, and almost everything people do depends on it.

Condition #3: It is active. When the curriculum is hands-on, collaborative, and fun, and when it helps students come to grips with high-level concepts, even reluctant learners tune in and learn. For example, Skyping with 16-year-olds in foreign countries, visiting a dim-sum restaurant or doing a treasure hunt in the city's historical society.

Condition #4: It stretches us. Students appreciate being pushed to their limits. That is what they experience in computer games. One student said of her teachers, "They see what you are like, able to be, and they just make it so much bigger." Another student said, "When I am challenged the perfect amount, I just wanna keep repeating the process. I wanna make it something I am great at."

Condition #5: We have a coach. "Students said they felt most motivated by teachers who acted like coaches in the classroom," states Cushman, "demonstrating new skills, providing support and encouragement, and helping them learn from their mistakes." Going over quizzes and tests is an ideal forum for coaching, getting students to figure out where they went wrong and fix their errors and misconceptions.

Condition #6: We have to use it. This can be tutoring struggling classmates, engaging in a mock trial in a history class, or sharing new insights on nutrition with family members.

Condition #7: We think back on it. "In the rush to move on, teachers may forget to provide students with an opportunity to reflect on the work just concluded," says Cushman. "What was difficult for them, and how did they manage those challenges? What would they do differently next time? How did they grow?"

Condition #8: We plan our next steps. Students need to see life's connections in simple tasks like doing homework and major efforts like senior-year capstone projects. "With enough prior scaffolding in self-managing their activities," says Cushman, "students can treat the senior project as a culminating demonstration that they are ready for adult life."

Complicating the issues of learning and performance is the role of MOTIVATION, a measure of the desire to learn in any social or environmental situation. Inherent in any learning situation is a judgement of risk vs gain. Learning involves an investment in time and energy—the payoff in acquiring new information for a skill to warrant the investment.

Motivation comes in the form of internal or external motivation.

Internal motivation appears not to involve external pressures for success, rather an intense internal need to learn and achieve. Examples include experts in any trade, professional and Olympic athletes, large corporation CEOs, academic leaders.

External sources of motivation include coercion, peer, parental, and family pressures, etc. It is likely that some internal motivations are the result of external pressures exerted over many years of living up to someone's expectations.

Regardless of the source, motivation is a powerful factor in learning and performing. There is a long history of research showing that motivation can be altered by adjusting a reward or a punishment. The fear of failure is a particularly powerful motivator for action or inaction. People are more willing to learn and perform for a large payoff than for a small one. Similarly, they will work harder to avoid a large punishment than a small one.

SMART, SAFE, PLAYFUL LEARNING

Louis Pasteur said, "Fortune favors the prepared mind." Approaches to learning that take the brain's connection to learning into consideration are safe and more prepared to be effective than those that do not. Safe approaches to learning are playful and release chemicals into our brain and body that support learning, while judgmental, unsafe approaches release chemicals that suppress learning.

Effective educators make students feel competent, safe, and smart. Perhaps there is no greater service any approach to learning, teaching, and performing could provide than to make students feel smart. That may seem like a narrow window, but it affords large rewards. Emotions, how students feel and see themselves, provide the starting points for a journey of meaningful learning and development.

SMART, SAFE, PLAYFUL environments are brain-compatible for learning, where making progress is more about anticipating and predicting outcomes than judgments of results.

S.M.A.R.T. CLASSROOMS = **S**tudents' **M**inds **A**re **R**eally **T**alented™

In S.M.A.R.T. classrooms, no student is seen as broken and in need of fixing. Unworkable outcomes are not judged, criticized, or seen as failures, but are viewed as valuable feedback for future reference. In these environments, teachers realize that every student is on a journey of development and his present skills and information base are exactly where they should be; nothing is missing; it is just developing. These developing students should be producing unworkable outcomes when learning.

The numbers we know before entering school (0 to 10) are the same numbers that build rocket ships; over time, we just learned to do something advanced and different with them. The same is true about the letters of the alphabet.

In the newspaper, *Huffington Post* (3/6/14), researcher David Hassell talked about stages of development and the importance of keeping individuals engaged. He pointed out that feedback is an important element for keeping students engaged and at their own rate of development.

He wrote that when given feedback about a weakness, individuals stay more engaged than when receiving no feedback. However, when the feedback reminds individuals of their strengths, on average, they are 30 times more engaged than those who are being told about weaknesses.

> **Approaches to learning that are trying to fix unworkable outcomes take note; mindsets that are trying to fix have never been a component of meaningful learning.**

S.A.F.E. CLASSROOMS = **S**tudents **A**lways **F**irst **E**nvironment ™

There are learning environments where students are told what's wrong, and they will not feel safe. This approach is only a negative reaction to unworkable outcomes and is not a proactive, safe learning experience that promotes an individual's self-confidence. After saying 2+2 is 5, many students hear, "wrong answer," instead of being asked "how did you arrive at that answer," and then, "how would you create that unworkable outcome on purpose."

A student-centered/brain-compatible information-delivery system recognizes that the main job of any instructor is to make students feel capable and self-confident while finding value in unworkable outcomes. Students are not failing; they are just experiencing a learning delay.

P.L.A.Y.F.U.L. CLASSROOMS =

Powerful **L**earning **A**bout **Y**ourself **F**inds **U**seful **L**earning™

"The highest form of research is essentially play." (N.V. Scarfe)

"We play to learn; we did not have to learn to play." (Michael Hebron)

Progress and improvement are grounded in the self, found in self-reliance, which is one of the many innate skills humans have when learning is based on play. We clearly play to learn; we do not have to learn to play. Trying this or that in a journey of learning and developing is brain-compatible.

The Gamification of Learning and Instruction, by Karl M. Kapp, is based on solid academic research over dozens of years by many studies that highlight the value of play in the cognitive domains of declarative, conceptual, and procedural learning and knowledge.

"What has been uncovered about the brain requires a fundamental change in education," Professor Harry Chugani, American Board of Neurology.

"We do not stop playing because we grow old; we grow old because we stop playing." (George Bernard Shaw)

The design of methods and strategies used for learning are often taken for granted and are undervalued. When compatible with the nature of learning, they can reduce frustration and intimidation by increasing playful brain-based learning opportunities.

"Play is our brain's favorite way of learning." (Diane Ackerman, author of *The Human Age, The World Shaped by Us*)

RELATE, REMEMBER, REPEAT

Relate, Remember, Repeat are three concepts that the nature of learning and the brain's connection to that process of change make use of.

In order to remember and repeat new information, the receiver of information (student, employee, player, child, etc.) must first be able to relate to what is being shared. The method of sharing must be provided in a manner that connects new information to what the receiver already knows or has experienced.

> ***To support a student's ability to relate to new information, use metaphors to introduce new information, insights, and concepts. I often tell students there is no new learning, what you are learning is a mix of prior knowledge with new insights.***

My example as a golf instructor would be expressing to the student, "You are not learning balance, timing or rhythm; you already use those skills in your daily life, and you are just bringing them to golf." Then I may say, "What you are learning will be like this or like that" depending on what is being discussed.

The aim is to avoid telling students what to feel, ("this motion will feel like this or like that"). Feel is too personal; one person has no idea what something may feel like to another person. Asking a question with the hope that it will help students relate to what is being shared. "Tell me what does this feel like to you? Does it feel like this or like that?" Again, give students a tool for relating new to old.

New learning always takes place on a stage of priors. Without RELATE, you will not have REPEAT or REMEMBER.

Acts of Coaching should join Acts of Teaching

When learning, the information delivered can be music to our ears, but can also be noise to our brain when delivered without the brain in mind. Teaching-fixing to get it right approaches are aimed at the present, and student-centered/brain-compatible, learning-developing approaches are coaching for the future. These approaches recognize the brain's internal operations must be taken into consideration first before the information is delivered.

I suggest that providers of information, educators, instructors, students, parents, or employers see themselves as COACHES for several reasons.

"Coaching is as much about the way things are done as about what is done. The means and style of communication used" (John Whitmore, *Coaching for Performance*). He went on to say "Coaching is not merely a technique to be wheeled out. It is a way of thinking, a way of managing, a way of treating people, a way of being. Coaching is not teaching at all, but is about creating the conditions for learning and growing, embarking on a personal development journeying." (Sounds brain-compatible to me!)

"Coaching puts attention on future possibilities, not past mistakes. Coaching is unlocking people's potential to maximize their own potential. Coaching requires expertise in communication skills, not only in the subject at hand. Coaching must think of people in terms of their potential, not their performance. The underlying and ever present goal of coaching is building the self-belief of others, regardless of the content of the task or issue."

Now read Whitmore's thoughts again and replace the word Coaching with Teaching, and you may recognize why I suggested all instructors should see themselves as COACHES.

David Rock, author of *Your Brain at Work*, launched Results Coaching Systems in 1998 and became a co-founder of the coaching certificate programs at New York University in 2005, suggests applying neuroscience discoveries to coaching. He wants people to make practical use of what is known about the brain. For example, doing drills to learn a new physical motion has been found to be less useful than doing drills after something has been learned.

Developing skills for understanding

Some of the following thoughts are adapted from Bill Marvin's insights as a business management consultant.

Some models of management are closer to a law enforcement approach than to anything else. "Find things that are wrong and fix them!" This is fine if you want to spend your life looking for things that are wrong. However, there is a more productive way to make progress.

Efficient coaching is an inside-out education, where the players see their own an-

swers. In contrast, what we saw in many schools was outside-in education where the right answers were held by the teacher.

To improve learning skills, get good at asking questions. Effective educators ask insightful, probing questions that cause students to think. It is hard to get yourself in trouble if you are either asking or answering questions. It is only when approaches to learning make statements (preaching or lecturing) that dangerous grounds are trodden upon.

Student-centered/brain-compatible, information-delivery systems provide a message of inclusion; they join acts of coaching with the nature of learning.

Traditional teaching tends to pay attention to subject content, rather than how we are going to use the information in a context of ever changing circumstances. Acquiring information, categorizing it, applying it, and reacting to outcomes involve our conscious and non-conscious minds in student-centered/brain-compatible learning environments.

THE BRAIN

The value of discussing the brain's connection to learning needs no defense. Without internal mind-brain activity, what could humans have accomplished, including meaningful learning? It has already been said, a master of anything was first a master of learning, and that is an accurate description of what the human brain is predisposed to do.

The following was sent to me by Dr. Stephen Yazulla:

"Any discussion of Brain-Compatible Learning should introduce the major players of the brain and nervous system, along with the environmental conditions in which the normal, healthy brain functions."

Development and Learning

Learning is like development in that learning also involves and requires changes in the synaptic organization of the brain. The changes that occur with learning are not as rapid or extensive as that occurring during development. However, the ability of synaptic changes to occur in both conditions is influenced by the chemical composition of fluids bathing the neurons. The chemical medium of the brain is influenced by what we eat, drink, breathe and put on our skin, as well as by our physical and social environments. The public and medical communities are aware of the great attention paid to factors affecting the brain as they relate to prenatal care and fetal development. However, there is growing attention by researchers to these factors as they relate to any Learning Environment.

A learning situation that takes effects on the brain into account to maximize learning is referred to as "Brain-Compatible." The brain is not a "Black Box" that is to be filled or tapped at will without considering the effects that the teaching methods or learning environments have on the brain. Stress, fatigue, diet, experiences, expectations are among the numerous factors that can affect the chemical environment of the brain during learning. Understanding the source of these factors to minimize negative effects on learning promotes a "Brain-Compatible Approach."

The brain and the spinal cord together comprise the central nervous system (CNS). All other nerves of the body, including those of the sense organs, muscles, glands, and internal organs form the peripheral nervous system (PNS). The human brain weighs about one pound at birth and reaches its final weight of about three pounds by 10-12 years of age. During the rapid growth phase in infancy, the brain adds about 250,000 nerve cells per minute by cell division (mitosis).

The brain contains two major types of cells: neurons and glia. Neurons (nerve cells) transmit, integrate, store and act on environmental stimuli. Input from sense organs, decision making, memory, and output to muscles and internal organs all rely on neurons.

Glia (from Latin, meaning glue) are supportive cells that surround neurons and are responsible for maintaining the chemical environment of neurons. Glia control the balance of water, salts, sugars, and other chemicals that modulate the activity of neurons.

When one thinks about the brain, one often thinks of the 'old gray matter,' i.e., neurons. Amazingly, neurons account for only 10% of the cells in the brain; 90% of cells in the brain are glia. Even though neurons are outnumbered 10:1 by glia, neurons occupy 40% of the brain volume.

Along with an extensive network of blood vessels, glia provide the critical function of house-keeping: energy supply, waste removal, growth, and repair; they protect neurons from substances in the blood (Blood-Brain Barrier), respond to infections, inflammation, and so on.

Neurons are specialized cells that transmit information by chemical and electrical signals. The input to the neuron ordinarily occurs at branched structures called dendrites. The output is transmitted along a "wire-like structure" called an axon by an electrical signal. Neurons communicate with each other by releasing chemicals at structures called synapses. The chemical travels a very small distance (about one millionth of an inch) to stimulate the dendrite of the next neuron in the sequence.

Chemical Messengers

Any single neuron can make synapses with only a few, or many thousands of neurons. Chemicals released at synapses are called neurotransmitters, and they act locally over very short distances, less than one-ten-thousandth of an inch. Some chemicals make neurons more active (excitation) while others make neurons less active (inhibition). The whole nervous system operates by a delicate balance between excitation and inhibition. The nervous system is never zero activity. It is always in a steady state that can be excited to do more, or inhibited to do less.

For example, motor neurons in the spinal cord that stimulate skeletal muscles (like in the arm or leg) receive a steady stream of excitatory and inhibitory input from the brain. Thus, at rest, skeletal muscles are in a constant state of tone. Even when relaxed, your muscles are under some tension; they resist pulling or stretching. In contrast, think of the dead weight your leg or arm feels like if it "falls asleep." Cutting off the blood supply to the leg or arm stops the chemical stimulation of the muscles, and they go limp. Similarly, the venom of a cobra blocks the excitatory chemical stimulation of the skeletal muscles, particularly those of the diaphragm that control breathing. The result is paralysis and suffocation. Whereas, a rat poison (strychnine) blocks the inhibitory input from the brain to spinal cord motor neurons. The result is that there is excessive and unregulated excitation of skeletal muscles, leading to convulsions and death. In each case, the balance of excitation and inhibition to skeletal muscles was upset, resulting in either flaccid paralysis (cobra venom) or spastic paralysis (rat poison).

Neurons, glia, and endocrine gland cells also release chemicals that can affect cells throughout the body. These chemicals are called hormones or neurohormones. Unlike neurotransmitters that act over very short distances between two neurons, these hormones enter the bloodstream and travel throughout the body. They act over large distances, affecting neurons and cells throughout the brain and organs in the body.

For example, if you are startled or frightened, you freeze, may feel a flush, a rush of heightened attention and alertness. This 'fight-or-flight' reaction is caused by a release of adrenalin and nor-adrenalin from cells in the brain and adrenal gland. The effect of these hormones is to increase lung and heart activity, blood flow to skeletal muscles, mental alertness, and attention to whatever caused the startle response. At the same time, there is suppression of the entire digestive system. This is one reason that if you engage in strenuous activity after eating it can cause indigestion. There is not enough blood in your body for all of your organs. So, blood that would be needed for digestion is diverted to your skeletal muscles, and the result is that the newly eaten food just 'sits in your stomach,' causing gastric distress. Adrenaline and other stress hormones, like cortisone, have enormous effects on the brain and cardiovascular activity. These stress hormones affect attention, alertness, memory, sense of well-being, all of which are important factors in learning situations.

Protection of the Brain Space

The adult human brain has about 100 billion neurons and one trillion (1,000,000,000,000) glia. Mixed among all these cells is an extensive web of blood vessels that supply nutrients (oxygen, glucose, amino acids, minerals, vitamins) and carry away waste (carbon dioxide, ammonia). In a normal condition, blood cells and neurons never mix. Blood cells are deadly to neurons. Head trauma (concussions) and strokes in which blood leaks into the brain can cause rapid death of neurons. Blood and neurons are kept separate by an elaborate system called the

"Blood-Brain Barrier."

For example, two cans (24 oz.) of a cola contain about 78 grams of sugar (3 ounces, 18 teaspoons). Shortly after drinking the cola, your blood sugar will show a rapid increase, but the sugar content of your brain environment will change very little. The "Blood-Brain Barrier" controls the concentration of sugar, salt and other substances that dissolve in water within very narrow limits to protect the brain from changes that occur during the day. However, substances that dissolve in fats and oil, for example, nicotine, cocaine, or general anesthetics, pass right through the "Blood-Brain Barrier" with potentially great effects on brain development and activity.

Brain Regions have different functions

The 100 billion neurons are not all the same; they are of different types, and they are not uniformly distributed throughout the brain. The idea that specific areas of the brain are responsible for specific functions is well understood today, but only has been accepted for the last 150 years. Franz Joseph Gall (1758-1828) made the first attempt to study brain function which is described in his paper "The Anatomy and Physiology of the Nervous System in General, and of the Brain in Particular." Gall's idea was that enhanced faculties of the brain would be expressed as expansions of the brain and would be visible as protrusions or bumps on the surface of the skull. This gave rise to the popular rage of Phrenology. Ceramic busts, with these brain faculties indicated, were popular and could be purchased at any Pharmacy. Although long discredited, Gall's efforts drew attention to the ideas that different areas of the brain were responsible for specific functions.

With advances in medicine in the 19th century, more people survived brain injuries, whether due to disease, accidents, or war. These provided opportunities to study the effect of brain injuries on behavior and thereby infer the area of the brain involved in particular functions. For example, in the 1860s language areas of the brain in the left hemisphere were identified by autopsy of patients who had severe problems expressing language (Broca's aphasia) or understanding language (Wernicke's aphasia). The situation regarding language is more complicated than indicated by these early studies. Still, the stage was set for the study of the functional anatomy of the brain, that is, brain and behavior.

Sir David Ferrier (1843-1928) published The Functions of the Brain in 1876. Over time studies into the brain would become less general, with specific topics including the brain's connection to learning being researched.

The brain has two hemispheres, right and left, that appear identical to any casual observer. However, functionally, they are different. In general, the left hemisphere controls sensory inputs and motor outputs for the right side of the body. The reverse is true for the right hemisphere. It is common knowledge that someone who has had a stroke in the left hemisphere is affected on the right side of the body, usually with some degree of paralysis. The two hemispheres are connected by a large band of nerve fibers called the Corpus Callosum that allows the two hemispheres to communicate with each other. In this way, the brain functions seamlessly as an integrated unit instead of separate structures.

The outer, visible part of the brain is the highly convoluted cerebral cortex. It is by use of the cerebral cortex that we think, reflect, and communicate with language and abstract symbols – functions that make us human. Other mammals have a cerebral cortex, but there is a qualitative difference between us that separates humans from them. What this difference is we can only speculate; but, it is not important for our purpose here.

There are functional differences among brain areas in the cortex. The visual sense is coded largely in the back of the brain; hearing on the sides, just above the ears; language ability, both understanding, and expression, is on the left side; skin senses and skeletal muscle control – half way between the front and back of brain; planning, intention, initiation of action, control of impulsive behavior (so-called executive functions) – in the front of the brain.

There are several specialized structures covered by the cerebral cortex that are important for our discussion of learning:

Hippocampus – is critical for the formation of memories.

Amygdala – is critical to assess the threat level or lack thereof of any incoming stimulus. The filter for the initial 'flush' before a threat is identified.

Basal ganglia – for coordination of skeletal motor control. Parkinson's disease is a result of damage to the basal ganglia.

Hypothalamus – thirst, hunger, body temperature; controls hormones involved in metabolic systems, growth, reproduction, immune system, kidney function and so on. Controls the pituitary gland and is really the "Master Gland."

Limbic system – is an ancient part of the brain that is involved in "primitive" attributes of emotion, rage, and sex.

The hindbrain connects the spinal cord with the rest of the brain; it includes Cerebellum–critical for fine motor control and what is called "motor memory" Medulla–controls breathing rate, heart rate and other "vegetative" functions.

As you might imagine, specific behaviors are correlated with increased activity in different brain regions as determined by functional Magnetic Resonance Imaging (fMRI). When you speak, the language areas on the left side of the cerebral cortex are more active; but when you are reading aloud, the visual areas in the back of the brain become more active. If you are walking at the same time, then the motor cortex on both sides of the brain get involved. The

Frontal Cortex has been directing this activity all the time, as well as the cerebellum that coordinates the automatic behaviors of balance and so on. As you are walking, your environment is evaluated for potential threats by the Amygdala and responses to these by the Limbic System and Hypothalamus.

The Changing Brain

As mentioned, the brain triples in weight from birth to adulthood, adding tens of billions of neurons during this time. All of these new neurons must be in the right region of the brain and connect with the appropriate targets. Chemical signals guide the rate of growth and direction of growth for the dendrites and axons of these neurons. Two examples below illustrate this important point.

Axons of sensory neurons in the hand must find their way to synapse on the proper neurons in the spinal cord, and these neurons from the spinal then grow into the brain. For movement to occur, the neurons in the brain must find their way into the spinal cord, whose motor neurons must go to the proper muscle to be activated.

Neurons in the eye must respond to the proper stimulus in space, find their way into the proper part of the brain to form an organized map of the visual world. These neurons in the brain must respond to the appropriate shape, size, color, and position of a stimulus, such that the person can say not only ‘I see it’ but ‘what it is,’ ‘where is it,’ ‘new or familiar,’ and ‘friend or foe.’

Even though the growing brain adds neurons, not all the neurons present in a child’s brain survive into adulthood. Why? Simply stated, the new and developing neurons are like small seedlings. As they grow, they sprout dendritic branches, increasing greatly in number, complexity, and volume. These branches compete for the limited space in the growing brain. Neurons that are activated together will form strong synaptic connections and survive. Neurons that are not activated by stimuli that are close in time or space do not strengthen synaptic connections, and these neurons eventually die. This gives rise to the well-known phrase in developmental neuroscience ‘Neurons that fire together, wire together.’ Inactivity also can weaken existing synaptic connections–‘Use it or lose it.’

Sensitivity of the developing brain

For the sensory systems of touch and vision, there is a critical time in development for stimulation to form the proper neuronal connections in the brain. For example, babies born with cataracts (clouding of the lens in the eye) will be permanently visually impaired if removal of the cataract is delayed until late childhood. Also, notice how easily children learn the language and how difficult it is to learn an additional language as an adult. This is because the developing brain is more dynamic “plastic” at a young age and becomes more hard wired as we get older.

The developing brain is very sensitive to its chemical environment. During pregnancy, consumption of alcohol, cocaine, amphetamine, nicotine, etc., by the mother can have serious effects on brain development of the fetus. This is because all these drugs get past the "blood-brain barrier" and have the same effects on the fetus as they do on the mother. Each of these drugs affects specific chemical transmitter systems and metabolism in the brain. The proper timing and coordinated operation of these transmitters are critical for proper brain development."

(Dr. Stephen Yazulla)

RECAP

Meaningful learning has been encoded when what is learned in one environment can be applied to a variety of dissimilar situations.

The components of meaningful learning include emotional, social, cognitive, physical, and reflective components. All five components function simultaneously.

Brain Compatible Approaches to learning are not fads or just other educational experiments. They represent meaningful changes from traditional approaches to learning and teaching based on the brain's connection to learning.

Student-centered/brain-compatible learning environments turn learning into pleasure, and pleasure into learning, with an emotionally safe process.

Metacognition is a set of five high-level brain controls (planning, selecting, connecting, expanding, monitoring) used from an early age that guide each individual's information processing; it plays a role in the process of learning and problem solving. These skills are enhanced each time they are utilized and will separate successful puzzle solvers from unsuccessful learners.

Safe approaches to learning are playful. They release chemicals into our brain and body that support learning, while judgmental, unsafe approaches release chemicals that suppress learning.

The brain is a highly complex structure containing millions of neurons and supporting cells. Informaiton is conveyed by numerous chemicals. The chemicals environment of the brain is affected by external events such as diet, stress, disease, joy, rage, grief, etc. The effects on learning and performing depend on our responses to these external events.

YOUR NOTES

CHAPTER ONE

Opening Insights

A century and a half ago (1879) Brook Adams, a member of the Boston School Board wrote, "Knowing a child cannot be taught everything, it is best to teach them how to learn." Over 135 years ago, Adams noted that most schools had no connection to such an approach. That is a statement that often rings true today.

After reading this chapter, one could open this book to almost any page as if it were one of the author's many notebooks containing useful insights for improving learning, teaching, and performing.

Studies about the implications of the brain's connection to learning are eye opening and more straightforward and understandable than one might anticipate.

Simply stated, the aim of approaches to education is to enable long-term learning which in turn is about change in both the brain and external outcomes. Consistent with my observations as an instructor, cognitive research has found that in all learning environments, including homes, schools, universities, business training programs and sports instruction, more individuals do not reach their full potential than do. Researchers contend that much of the information shared with students is not being learned. This is unacceptable, considering our so-called technological advantages and what we know about the brain and its connection to learning.

Are students learning what we think they are learning?

The goal: better learning and a better way of getting there.

The goal: to move current issues and future promise in the direction of breaking down information into digestible concepts that can be used.

The goal: that everything done during a learning opportunity is done to influence the future.

Albert Shanker, past president of The American Federation of Teachers, once estimated that only one-fifth of American students are well served by traditional classrooms. In 2013, according to Department of Education figures, only 32 percent of high school graduates are prepared for college. As per data reported for the ACT and SAT college entrance exams, the situation did not improve much in 2015.

Instructors and students are individuals drawn together as a team with goals and responsibilities.

The student's goal is to learn.

The educator's responsibility is to support the student's goal by using an appropriate process for sharing, thus enabling the student's ability to learn information and skills.

The student's responsibility is to participate actively in the learning experience.

The instructor's goal is to help the student learn to think critically, and apply learned content to new situations.

Studies at Columbia University Teacher's College and Harvard University's Graduate School of Education have shown that the design and structure of "the approach" to learn which individuals are exposed to will influence their pace of progress in school, sports, and other endeavors as well as their future employment opportunities, living conditions, physical and mental well-being.

Elizabeth Green, Director of Doctor of Education Leadership, Harvard Graduate School of Education (New York Times, 3/17/14), "Among the factors that do not predict whether a teacher will succeed include: a graduate school degree, high scores on SAT, extroverted personality, politeness, confidence, enthusiasm or having passed the teacher certification exam on the first try." That is quite a list of non-starters!

> ***Perhaps the attitude instilled in aspiring teachers is a key, as stated by Ms. Guzman, of the Lawrence High School Learning Center, Lawrence, MA, "Instead of focusing on what students are doing wrong, focus on what they are doing well, and that kind of pushes people." (Education Week, 3/6/2015)***

"People want answers to questions about memory, consciousness, learning, free will, and emotions." (D.F. Swaab, *We Are Our Brains*) While research into learning is constantly increasing, what has already been uncovered has made a positive impact on improving and enhancing strategies for learning when put to use.

Nora Newcombe of Temple University's Spatial Intelligence and Learning Center states, "Teacher training should include the most recent developments in cognitive science; in many teacher-training programs, the information being used is 40 years out of date." (Scientific American, 9/2014)

> Homo sapiens means "wise man;" so, what is the problem?
>
> Why do some individuals experience meaningful learning while others do not?
>
> What should we know about the nature of learning?
>
> How should students be evaluated, if at all?
>
> Is there a need to rethink the relationship between learning and teaching?
>
> How often do we consider that what goes on in the brain influences our ability to learn, and everything else we do?

Is the brain so much a part of our lives that we take all it does for granted?

When students are not making progress at an acceptable pace, do they need more education, or do the methods and strategies that the approach to learning uses need to be revisited?

Those are the kind questions that started me and many others on a journey in the early 1990s (the Decade of the Brain) of becoming more informed about how the brain organizes itself, and in turn, organizes both the physical and mental outcomes we experience when learning.

During learning opportunities, I began to realize that the information being shared is only one side of the story. The process(es) the instructor uses to deliver the subject content is the other, more important side of the story. All messages require a medium and some messages are more compatible with the nature of learning than others. One of the messages here is that how information is shared with receivers matters; it influences things to come.

Learning with the Brain in Mind is a call for what many believe is a much-needed change; a step away from poor assumptions and current fault lines found in some approaches to learning. **It is a move in the direction of incorporating the dual roles of delivering accurate information in emotionally safe, student-centered/brain-compatible learning environments.**

Experiencing meaningful learning is a process that is more than just sharing and receiving accurate information. More specifically, it is also about taking into consideration how cognitive development and learning are connected. What follows will travel inside the process of learning and change, offering research and insights that support getting off to a good start and making progress during learning opportunities. Studies about the implications of the brain's connection to learning are eye-opening and more straightforward and understandable than one might anticipate.

Perhaps some may view what has been compiled here about ways of enhancing approaches to the process of learning as a criticism of current methods. That is not the case; the only intent is to share insights into the brain's connection to learning that readers may not be aware of. Some of these insights, though common among academic researchers, are now appearing in the popular press more frequently.

We do not have to try to learn; we just can and do learn in a student-centered/brain-compatible learning environment. Any provider or receiver of information who gains an understanding of this connection will have an advantage. Fortunately, what science has uncovered about learning can be used immediately. **No one is born a poor learner, but helplessness learned through persistently frustrating experiences can cause people to simply stop trying.**

There are books about learning methodology for audiences that include educators, parents, employers, coaches, designers of curricula, instructors, business training teachers, etc. There are also books for students who want to become more informed about acts of learning. This book was written with the aim that it would contribute to a basic understanding of the nature of learning and assist both groups in becoming what University of Washington's John D. Bransford calls, "self-sustaining, life-long learners."

In 1899, the famous American psychologist William James published a little book, *Talks to Teachers*, in which he sought to explain how to apply psychology to education—that is, he sought to use what he called "the science of the mind's workings to generate practical advice for classroom teachers." (Ambrose, et al., *How Learning Works*, 2010)

Following William James' lead, it occurred to me that current studies gathering insights about learning could be applied to learning any topic in school, business, and other learning situations, including learning and improving sports skills.

Learning with the Brain in Mind is organized in a series of chapters about the active process of doing learning. These chapters are based on studies showing that the approach to learning should take the brain's internal operations into consideration before providing information to be learned, remembered, and used. It is a pluralistic approach that combines information with the processes of providing that information. It is a process that has a unity of thought about the nature of learning that supports the diversity that exists in our student population.

When referring to all our natural or innate behaviors such as balance, strength, and rhythm, our brain's natural or innate ability to learn often gets overlooked. One of the most valuable achievements of modern science is the uncovering of insights into the process of change. This connection is referred to by Todd Maddox, (University of Texas, Austin), as "our brain's learning system" which is discussed here.

"The great American novel, *The Adventures of Huckleberry Finn*, is about a boy who hates school. *The Adventures of Tom Sawyer* is about a boy who hates school. *Catcher in the Rye* is about a boy who hates school, and *A Separate Peace* is about a boy who hates school as are scores of other books. Schools seem to be an almost universally unpleasant experience." (Joe Queenan, *The Wall Street Journal Review*, 5/23-24/2015) ***But WHY? What could change and be different?***

It has been said that continual learning and developing are keys to life's success, but many of us know little about how meaningful learning takes place. Authors of *Making Connections: Teaching and the Human Brain,* Caine and Caine, pointed out "How individuals learn, is often never examined by a vast number of educators or the industry." This view has started to change for the better!

Professor Terry Doyle, author of *Learner Centered Teaching: Putting the Research on Learning into Practice*, "I think in the past ten years there has been a significant improvement in teacher's understanding of learning. I believe we have a long way to go, but I think progress is being made. I always say it is not possible to design learning activities and environments that facilitate students learning if you do not understand how students take-in, process, and retrieve information. We must know how learning happens if we are to be effective teachers."

If you ask the following questions about the theory of this book, the answer to all questions is yes:

> Does the theory reflect the real world of human beings?
>
> Is the theory supported by convincing evidence?

Does the theory explain past, present, and future outcomes?

Is the theory open to new research?

Is the theory understandable?

Is the theory emotionally self-satisfying?

"A good theory for anything should account for everything going on; if it does not, we have no idea what we are talking about, it should also have predictive power; if it does not—back to the drawing board." Prof. Angus Deaton, Princeton University.

> ***"Everyone has a brain ... If you want to understand why you feel the way you do, how you perceive the world, why you make mistakes, how you can be creative, why music and art are inspiring, you need to understand the brain." (Jeff Hawkins and Sandra Blakeslee, On Intelligence; 6 & 2)***

"The biggest mistake is the belief that intelligence is defined by intelligent behavior. The seat of intelligence is the brain's neocortex." "It is what somebody does in his or her brain and nervous system that produces a result." (Tan James, M.S., Ph.D.)

Information that any instructor shares or that students gather on their own is not as important as what is retained from learning opportunities. Methods and strategies that educators, parents, sports coaches, employers, or students use, strongly influence what is *retained* from any learning opportunity.

Acts of doing learning are influenced less by the challenge of subject content than by the culture of the approach to learning that individuals experience. Some approaches to learning are more purpose-driven than others as students discover what to learn, why to learn it, and then how to transfer what they learn in one context to other environments.

An important insight is that student-centered/brain-compatible learning environments are meant to be an investment in human capital; they enable the freedom for the student to discover and evolve their own learning. These approaches are not narrative-driven environments as they are influenced mainly by the process of learning.

They support thinking smart, not more thinking. They support hearing what is not said, seeing what is not there, and having answers to questions not yet asked. This book is a Learning 101 discussion founded on basic information about the implications of the brain's connection to change; a journey each individual must continue to pursue to evolve their full learning capacity.

If we expect to be fully informed about gaining an education or helping others gain theirs, then approaches to learning should keep the brain in mind. It helps to realize that what goes on in the brain, consciously or non-consciously, influences outcomes when learning. From that point of view, *this book is about thinking*. It is a book about you and me, and every other provider and receiver of information in learning environments.

> ***Peter Kline points out in The Every Day Genius that "The brain is an instrument that can be used either well or poorly. It is not difficult to use it well, but many of the influences now in education condemn students to use their brains less efficiently than they could."***

Robert Sylvester, professor of Education, University of Oregon, pointed out that problems arise when students try to be reflexive before developing the reflexive skills that are necessary to solve problems in deliberate, creative ways. Reflexive skills prioritize the steps that we want to take when learning. The brain operates by making predictions, "if this, then that," a process supported or suppressed by the approach to learning that is being used.

Like, want, and need are often seen as different emotions, but when it comes to the nature of learning, these three terms apply.

We *like* to learn.

We *want* to learn.

We *need* to learn.

Learning with the Brain in Mind connects these three emotions with the aim that in the future, students will profit from what instructors are sharing.

STARTING OVER

At some point, everyone has been a teacher and a student. That said, what would change misconceptions about how we learn or deliver information to others? There are behavioral interventions that can provide a supportive starting point when one is informed about the brain's connection to learning.

Teachers can be:

> ***Energizers: Boost your spirits, create positive feelings; the person you call when you need a laugh.***

> ***Mind-openers: Expand your vision, introduce you to new ideas, and challenge conventional thinking.***

> ***Navigators: Share advice and direction; help you talk through options to find your path.***

So can students!

The book *How Learning Works* (Ambrose, Bridges, DiPietro, Lovett, & Norman, 2010) offers these descriptions of learning.

> 1. Learning is a process, not a product. However, because this process takes place in the mind, we can only infer that it has occurred from students' products or performances.
>
> 2. Learning involves a change in knowledge, beliefs, behaviors, or attitudes. This change unfolds over time; it is not fleeting but rather has a lasting impact on how students think and act.
>
> 3. Learning is not something done to students, but rather something students themselves do. It is the direct result of how students interpret and respond to their experiences—conscious and unconscious, past, and present.

Student-centered/brain-compatible information-delivery systems have a culture that recognizes that meaningful learning environments do more than just deliver information. The components include information, why it is delivered, when it is delivered, who delivers it, how and where it is delivered, all with the aim of advancing learning opportunities.

Respected educator and researcher John Dewey (1896-1936) believed schools and teachers should study learning.

Often there is a distinction between what educators do and what they could do to support learning.

Enjoy your journey through this book…

THE BRAIN IS THE GATEWAY TO LEARNING.

THE PROCESSES AND STRATEGIC APPROACHES

ARE THE GATEKEEPERS.

BEFORE CONTINUING

Before continuing this discussion on Learning with the Brain in Mind, let's start with a questionnaire.

Based on your current views, please evaluate the following statements as "mostly" true, or "mostly" false. The same questions are provided at the end of the book for you to compare your responses after completing the book. For this exercise to be most effective, please consider your response to be your current opinion, not a response aimed at passing a quiz!

1. Meaningful learning environments are social environments. T F

2. Having students learn details can be less useful than helping them develop approaches to learning. T F

3. We should try to fix unworkable outcomes when learning. T F

4. We learn best from whole concepts, patterns, and sequences, not details. T F

5. The non-conscious mind is not valuable when learning. T F

6. Long-term learning is encoded after a lesson, and often during sleep. T F

7. Expert models are best used as models to copy. T F

8. Praise can be punishment in learning environments. T F

9. A master of anything was first a master of learning. T F

10. Actions of parents, employers, coaches, and educators should make individuals who are learning feel capable. T F

11. Being taught what a perceived expert believes is correct is less useful than students developing their own view. T F

12. A structured approach to learning is more useful than a random approach. T F

13. Learning often makes unconscious shifts (aha! moment) from a belief to a fact. T F

14. When learning, both workable and unworkable outcomes have value; with workable outcomes, more useful. T F

15. New learning is encoded as the safety of the environment, and the emotional state of students are simultaneously evaluated by experiences stored in the brain. T F

16. Memory is more an act of rebuilding than recalling. T F

17. We did not have to learn to play; we play to learn. T F

18. When learning and performing, outcome goals are more useful than having learning goals. T F

19. The task of an instructor is to make students less dependent on them. T F

20. To be wrong is part of the process of gaining understanding. T F

21. It is better to be right, than curious with an open mind. T F

22. Knowing information is different from being able to use that information. T F

23. The human brain is wired to protect us from danger; physical and emotional. T F

24. Our prior ways of thinking do not impact new learning. T F

25. It is important that students avoid doing the wrong thing. T F

26. Positive thoughts improve attention. T F

27. Trying to suppress doing the wrong thing often recreates it. T F

28. Emotions can support or suppress learning and change. T F

29. Experience influences our thoughts in the now, and can stop or support learning and change. T F

30. The brain has an unlimited capacity for encoding information, but a limited capacity for recalling information. T F

31. Training and studying in one place for a period of time is more useful than several shorter time-frames and frequent change of locations. T F

32. Introducing errors when learning, supports experiencing meaningful learning. T F

33. The brain stores through similarities, but retrieves by differences. T F

34. It benefits learning more when notes are handwritten rather than typed. T F

35. The more information is distilled, broken down, and analyzed by instructors and students; the more it supports learning. T F

36. Intelligence comes before the behavior. T F

37. Information in the brain travels between neurons at over 200 mph. T F

38. Intelligence and understanding start as a memory system that feeds predictions into the sensory system. If this, then that. T F

39. Prediction, not behavior, is the basis of intelligence. T F

40. The brain self-learns; computers are programmed. A computer has to be perfect to work; the brain is flexible and tolerant of failures. T F

NOTE: Answers are in the Quiz Retake in Chapter 12.

This questionnaire is introduced at the beginning of this book following the lead of noted psychologist Dr. Elizabeth Bjork, of UCLA's Learning Lab. She found that giving students a quiz at the start of her course enhanced their ability to experience new learning. She pointed out that tests do not have to be seen as an evaluation of what students know, or do not know. Rather, they provide an introduction to what should be known after taking her class. A pre-test can quickly turn unfamiliar concepts into personal insights for future learning, and application.

Dr. Bjork's studies found that pre-testing served to prime the brain, predisposing it to absorb new information. She points out that a student's guess will link to future learning. It was found that pre-testing leads to activating, reshaping, and enriching a student's mental network, embedding once unfamiliar concepts. The brain networks that were active during a quiz are reactivated when information is reintroduced during future instruction.

CONTENT

This book discusses insights that providers and receivers of information may want to take into consideration during learning opportunities. These insights do not prescribe exactly how to teach but do help to predict how effective the learning opportunities can become. It has been demonstrated that by taking the various researched recommendations summarized here into consideration, approaches to learning became significantly more effective, and student learning improved.

The studies referred to point to many elements that influence learning including: the student's prior knowledge; their motivation, development, mental level; opportunities for practice; efficient feedback; and learning to become a self-directed learner. These studies support the main topic of this book; provide student-centered/brain-compatible learning environments that are playful and frustration free.

In our fast-paced, result-oriented culture, some approaches to learning have become more a product of sharing content than a process of investing in students. This is true not only in schools but also in business and sports training. Unfortunately, politics, some existing cultures, and personal bias as related to learning, can keep entrenched policies in place despite overwhelming evidence to the contrary. ***Unfortunately, there are methods of learning that have been found to be inaccurate, or even damaging, leaving behind uneducated students who are often incorrectly referred to as "poor learners."***

Putting politics aside, many believe it is time to exchange 20th century teaching for the 21st century learning with a contemporary learning process. In these environments, most of the responsibilities for learning shifts from instructors to students. In this model, the instructor's responsibility is to guide students in the direction of developing tools to learn with. Students are then responsible for putting those tools to use while learning from guided self-application and self-assessment.

It has been said that there is nothing more expensive than a closed mind. "Personal bias" is listed as one of the main reasons for students not learning new information and remembering in *The Seven Sins of Memory* by Daniel Schacter. Having an open mind when reading what has been compiled here about learning will add value to what may initially be perceived as non-traditional, and/or counterintuitive content.

Most of this book is written in a non-progressive style, and for the sake of simplicity.

To enhance methods of sharing information and create the best learning environment for the student involves knowledge of a flexible process. It involves a combination of both "emotionally safe" and "guided self-thinking to solutions" that support and encourage learning.

The aim of the information presented here is similar to the one Dartmouth College professors John Kemeny and Tom Kurtz had in 1964. With a team of eager students, they developed a computer program written in BASIC (Beginners All-Symbolic Instruction Code), a programming language designed for non-engineers and non-mathematicians. *Learning with the Brain in Mind* contains general information about learning directed at a broad audience of instructors and students.

This book discusses enhancing instruction delivery methods to enable:

Long-term learning

Talent development

Unlocking intelligence found in one's experiences

Understanding the benefits of workable and unworkable outcomes

A curriculum put forward by an expert, without the consideration of the delivery method, rarely improves the ability to reason toward workable outcomes in real world environments. Understanding the steps of putting to use new learning in changing environments is just as important as the subject information that individuals are receiving; perhaps more important. By incorporating some changes into the way information is delivered, meaningful long-term learning can become as easy as one, two, three.

1) Approaches to learning should be compatible with the BRAIN'S connection to learning from workable and unworkable outcomes.

2) Learning environments should be PLAYFUL, and emotionally and physically SAFE.

3) The approach to learning should use metaphors that mix a student's prior knowledge with new content.

To accomplish this may require re-evaluating personal assumptions about strategies and methods used by traditional "teaching-fixing to get it right" approaches to learning that can limit a student's ability to reach their full potential.

Several fields of study referred to as the "Learning Sciences," have uncovered insights into human development that have resulted in the production of a more meaningful framework for educational methods in which learning and performing can be optimized. For example, a basic pursuit of cognitive neuroscience is to understand how information is organized in the brain while cognitive psychology is directed more at how humans process information.

All behavior, including learning, is an expression of biological functions originating in our brain.

Acts of doing learning can be advanced by moving beyond the obvious elements of only taking in accurate information. Insights that support efficient learning are found in biological science and in particular in brain research. Maybe the term "learning" should begin with the letter B for "Biology of the Brain." The heart is a Biological pump; the nose is a Biological filter; the Brain is a Biological information processor.

Aristotle talked about what he called "final causes." Perhaps learning should be seen as the final cause of our species' survival.

DUAL ROLES

Many approaches to learning are at war with the obvious. Often there is a significant disconnect between what instructors know about a subject (accurate information) and their ability to create an environment where meaningful learning takes hold.

This is unlike some other professions, for example:

A lawyer's responsibility is to provide legal help.

A doctor's responsibility is to restore health.

A car mechanic's responsibility is to repair cars.

An educator's responsibility is to help individuals learn.

Lawyers, doctors, and car mechanics know both sides of the story. They are informed with subject content information, but also have the know-how skills to apply that information to accomplish their jobs in ever changing environments. They play dual roles of gathering knowledge and applying information.

On the other hand, some educators, coaches, and other instructors can be aware of the accurate subject content and, at the same time, be uninformed about taking into consideration the most effective manner of presenting the subject to the student.

Today, most of the information shared with receivers is already available on the internet. This makes doing learning less about information and more about how that information is shared.

Is the process being used compatible with the nature of learning?

QUESTIONS

In the fall 2010 issue of *Thought and Action*, a National Education Association Journal, it was written that every day after school Noble prize recipient Isidor Rabi's mother would ask him, "did you ask any good questions today?", not "what did you learn in school today?" His mother understood that the roots of all learning are circulated in several directions by a mindset that emphasizes the active process of questioning rather than the passive gathering of facts.

Recognizing the value of Rabi's mother's question, I hope that the questions and insights compiled in this book will prompt coaches/educators, and students or puzzle solvers, to ask their own good questions about the brain's connection to experiencing meaningful learning.

Questions support the brain's connection to learning (retaining the information for the long-term) when they cause thinking, rather than just receiving information (short term recollection) to provide the correct answer or result.

Many reflect on learning as if an unworkable outcome needs to be fixed. The aim of student-centered/brain-compatible approaches to learning is to let go of judgmental thinking and realize that a fluid physical function or new learning is based mostly on what can be learned from unworkable outcomes. Unworkable outcomes become our teacher of what to do differently. This is an important insight.

EMOTIONS

"Emotions underpin how students learn; they steer our thinking." (Mary Helen Immordino-Yang, professor of education, psychology, and neuroscience at University Southern California). I attended her lectures at Harvard's Connecting Mind Brain to Education Institute. In a new book, *Emotions, Learning and the Brain*, Yang and her colleagues discuss how emotions influence learning.

Learning and creating require that we are first thoughtful. "We have to connect the dots" is a popular saying. For new learning to take hold, new connections between neurons in the brain must occur. From a neurological perspective, we come into the world learning-ready. Learning happens as a result of how our thinking-doing mind responds emotionally to the positive and negative words, thoughts, and situations we encounter while living, learning, and performing.

> ***Do methods used by approaches to learning create emotions that motivate or do they create ones that suppress a student's ability to experience learning to the extent of their potential?***

Brain compatibility, when it comes to learning, also means emotionally compatible. Learning and emotions are always connected, an insight that should not be overlooked. Approaches to learning should act as caretakers of learning skills. Emotions are automatic responses to situations; responses that were learned from our own experiences. Emotions and learning could be seen as two sides of the same coin.

While the brain is central to learning, Martha Kaufeldt, author of *Begin with The Brain*, points out, "If the brain has to deal with frustration, fear or confusion, its performance is inhibited." Anxious students become poor learners.

Useful questions!

Does the approach to learning use methods that create intimidation and frustration?

What supports or suppresses memory and the recall of information and skills?

Does our method improve learning opportunities by decreasing or eliminating student frustration and intimidation?

Is the approach to learning geared for just sharing information, or for skill development, learning, and performing?

A Master of anything was first a Master of Learning.

By providing an emotionally safe, playful learning environment, a student's potential can be enhanced. The root of the word "school" is found in the Greek word "schole" meaning LEISURE.

Leslie Hart, the author of *Human Brain and Human Learning*, would ask: Was there an absence of threat from the approach to learning, simultaneously engaging the student's intellect, emotions, creativity, and whole body?

Experiencing meaningful learning should not be a tricky business. It should be emotionally safe and playful in nature. This view provides implications for designing approaches to learning.

Emotions are discussed more fully in Chapter 9.

LEARNING IS A PROCESS

"The path is the goal; the goal is the path." (Buddhist view on Learning)

"The process is the aim; the aim is the process." (Learning with the Brain in Mind)

There are approaches to learning using a process that releases chemicals into our nervous system that support the experience of meaningful learning, whereas other approaches release chemicals that suppress learning. When designing and delivering information to students, being informed about what suppresses learning is as important as knowing what supports learning. Experiencing meaningful learning is more than just knowing facts. Accurate information is less valuable when presented to students with a method that is not compatible with how students learn best. Methods normally have specific approaches and outcomes in mind, on the other hand, the process of learning is fluid and flexible.

It is the process of learning, rather than just subject content information, that strategies for teaching need to aligned with. If educators and students do not understand that the learning process is the chief enzyme of education, students are not learning as much as they could have.

"To improve learning, teach students how to learn." (Catharine Fosnot) This statement would apply equally to any provider of information.

In 1937, Heinz Werner drew our attention to the traditional emphasis on achievement (high grades), rather than on the process of learning. He accurately predicted the unfortunate consequences of unprepared students for real-world application of knowledge. In 1960, Jerome Bruner wrote *The Process of Education*, which rivaled Alfred Whitehead's *The Aims of Education*. Both books highlighted the process of learning and the value of having individuals discover knowledge for themselves.

"An expert in facts, as in the TV game Jeopardy, is useless, unless the information can be critically evaluated and integrated in some coherent useable way."(Dr. Stephen Yazulla)

The insights that Werner, Bruner, and Whitehead were sharing about the process of learning in the past are now supported in the 21st century by current research. As Malcolm Knowles said, "Learning is more efficient using a learning plan than a course outline."

Short thoughts on the process:

The mind unconsciously remembers when in the process

Freedom is not doing what you want; freedom is knowing who you are and simply and freely being in the process

Intelligence is usually defined by behavior; behavior is influenced by predictions; predictions are influenced by our experiences. Moreover, the non-conscious experience is what we use while in the process.

If you do not build it in the process, you will not understand it.

We are ourselves in the process, but results and outcomes are not who we are.

In the process, outcomes are because of the process.

The process can be damaged when we are not ourselves.

In the process, we let go of emotions after an unworkable outcome.

Memories without negative judgments are valuable to the process.

In the process, we learn to let workable outcomes arrive—without avoiding unworkable outcomes.

In the process, learning happens somewhere between not-knowing and knowing; between workable and unworkable.

In the process, we often do not know how we did it, but our unconscious does.

In the process–no expectations about outcomes–no conscious thinking.

In the process, we do not consciously add, subtract, or go from complex to simple.

Entering a learning environment can be a conscious act, but in the process, learning more often than not, is an unconscious result.

In the process, learning needs little else but being in the process to capture the reality that surrounds you.

When learning, be who you are and stay in the process.

In the process, the self that makes us who we are is not a solid form; it is fluid, flexible, and portable.

In the process, the self in yourself is what meaningful learning and creativity travel through.

The process flows through a trust in your mind, thoughts, and feelings.

When you go deep enough into the process, it will take the self in you where you need to go.

Deep cannot be found in the width of details, deep exists in personal insights uncovered in the process.

The process has no straight lines; its strength is found in randomness that lacks a map while supporting self-reliance.

In the process, there is no wrong or right way. All approaches are useful some of the time.

When the scared or insecure leave, suddenly unintended, raw energy arrives in the process.

In the process, inspiration means taking in what is happening or being felt.

In the process, it is fair to say a healthy brain is always operating on our behalf.

Within the process, we are adjusting to the environment, even though we are unaware of these internal changes.

In the process, learning is always alive with possibilities, not with answers.

Effective educators, guides, and coaches never tire of helping others to become aware of the process.

MOVING FORWARD

The Institute for Learning and Brain Science, University Washington states, "We believe that the discoveries about the brain's connection to learning will be comparable to those in genetics." Science progresses in steps from simple to complex, from one answer to another question, and so on. This is true with the ambitious effort devoted to human genetics and neuroscience.

The brain is not designed to fail tests, have car accidents, miss putts in golf or do poorly in business. Out of the estimated 8.7 million different living species on earth in the 21st century, 50 species become extinct every day. However, humans, thanks to our mind-brain, are still here developing, surviving, adapting, and learning. We do not have to try to learn; we just can and do it in appropriate student-centered/brain-compatible learning environments.

Gaining insights into the nature of learning and the brain's internal connection to this process is a journey in two directions at the same time. One direction is moving away from damaging approaches to learning, and the other is moving in the direction of experiencing long-term learning. Meaningful learning is supported by moving toward self-discovery and self-reliance skills and away from customs that foster methods that produce a poor self-image and lack of self-confidence.

Suggestion: *Design approaches to learning that inspire students and protect the kind of curiosity that supports meaningful learning.*

"Our brain, although, we do not see, touch, hear, or smell it, is still the most important part of us. It is the most human thing about us." (W. R. Klemm, Science, *The Brain and Our Future*; 2) The nature of learning is where curiosity, imagination, memory, and emotions reside in chemical-electrical messages among neurons that support doing learning. Unlike a computer, the brain can program itself, a process that does not change as circumstances or topics to be learned while experiencing change.

The end of the journey is not knowing more; it is doing more." (Julie Dirksen, author of *How People Learn*)

Francis Bacon, 400 years ago, said "Knowledge is power," but that is not exactly correct. How information is acted on and used is where power lies. How the information is acted on will always be influenced by how information is delivered and emotionally received during learning opportunities.

YOUR NOTES ____
____
____

CHAPTER TWO

Myths and Facts

There are many long-held myths about how we learn that can be more attractive than the informed realities they are overlooking.

"The mystery of life is not a problem to be solved, but a reality to be experienced. " (Art Van der Leeuw)

Some say it is irresponsible to ignore what is being learned about the brain; it helps us make better decisions in offering a quality educational experience.

Universities across the country are endorsing how we learn. Their programs now include a combination of both the content areas of learning (subjects) and the cognitive neuroscience of learning and their objectives.

COMMON LEARNING MYTHS

Many long-held views about learning are disappearing in the 21st century. Professor Daniel Willingham published research in 2005 that showed we should teach to the content of the subject, not to a learning style. Then seven years later, a 2012 Wall Street Journal article by Professors Christopher Chabris (Union College) and Daniel Simons (Univ. of Illinois) revisited the Learning Style Theory along with two other brain myths.

They began their article with three statements asking which one was false:

1. We only use 10% of our brain.

2. Overly stimulated environments will increase the intelligence of preschool children.

3. Individuals learn better when they receive information in their preferred learning style; those being visual, logical, verbal, physical, and aural (hearing) learning styles.

It turns out all three statements are <u>false!</u>

The article pointed out that 242 teachers took part in a study by Sanne Dekker and colleagues at the Universities of Amsterdam and Bristol.

Their findings reported that the most common and popular brain myth was about learning styles. Ninety-four percent (94%) of the teachers in the study believed they should teach to the student's learning style, a popular approach.

"One of the most accepted understandings of learning styles is that student learning styles fall into three "categories:" Visual Learners, Auditory Learners, and Kinesthetic Learners. These learning styles are found within educational theorist Neil Fleming's VARK model of Student Learning. VARK is an acronym that refers to the four types of learning styles: Visual, Auditory, Reading/Writing Preference, and Kinesthetic." (https://teach.com/what/teachers-teach/learning-styles/)

This attitude of teaching to a student's learning style puts a huge burden on teachers who must adjust to each student, assuming they can judge a student's preferred manner to receive and process information. Of course, in the real world, outside of the classroom, employers, colleagues, social contacts, etc., are not likely to cater to individual idiosyncrasies. Alternatively, unilateral teaching to modality puts the burden on the student to adjust to a common manner of teaching a subject matter regardless of what is believed to be their preferred style.

The Association for Psychological Science found that there is essentially NO evidence that customizing instruction to match a student's preferred learning style will lead to better achievement. **If you want students to see something; or hear something; or feel something, teach for that outcome, not to a preferred learning style.**

The Chabris/Simons research also found that "47 % of the teachers believed that we only use 10% of our brain while contemporary studies show that to be incorrect; **we use the entire brain**."

Also, 76 % of study participants believed that exposure to a Baby Einstein™ type video to be enrichment and that going beyond what is already a significant developmental environment would improve a child's cognitive development—contemporary studies show this is not true.

Why do people believe in a theory that has no cross-research behind it? They believe it because it fits a general assumption; or because others believe the theory; or because it is perceived to have become common knowledge; or because of extensive media exposure promoting it.

Topic: Building a Better Model for Learning to Replace the Myths That Exist. The following notes are based on a Kurt Fischer lecture, director of Harvard's University's Connecting Mind-Brain to Education Institute at Harvard.

Some myths Include:

Males and females have fundamentally different brains.

Knowledge is something that we give to each other.

Learning is about filling the brain with information.

There are left-brained, and there are right-brain learners.

Some facts include:

Train the brain to do what it wants to do—learn efficiently.

Learning is based on what has been experienced.

Developing and learning are not linear processes. Learning is a web of connections in the brain. The brain operates like a committee, with most members having their say all at the same time.

In less than a second after we see a word, language, emotions, etc., just about everything comes into play.

The brain is a parallel processor. Simultaneously, the brain evaluates incoming information with stored experiences to determine what is safe or unsafe; useful or not useful, etc. This occurs before the information is encoded in long-term memory.

Things that affect our learning include anger, fear, emotional evaluation, and our perceptions.

The brain wants to act efficiently, using the smallest amount of energy it can. Unskilled individuals use more brain activity, which is less efficient.

We self-build and construct knowledge when learning. Learning is a dynamic, active process that takes time. Learning is developmental.

Learning happens in networks throughout the brain, during emotional evaluation.

We learn through a history of questions. We grasp and build with our mind's experiences.

The whole brain resonates with robust learning through different brain processes.

Research informs practice and practice informs research.

Often good courses can produce poor learning, with little evidence of understanding, problem solving, or learning.

Use implicit models (personal models, non-external) and metaphors.

It is our experiences that hold together understanding.

When learning, it is useful to know that functions can improve to optimal, but will fall off and can return. The gap is larger for adults. We regress and then rebuild.

We learn best in safe, playful domains and in context.

Unfortunately, fast learning produces rigid forms and inflexible thinking.

Over time, slow learning produces more variables and flexibility.

Our actions build representations and abstractions within the networks of the brain. The body contributes to learning, providing information that shapes learning.

Additional myths versus facts about learning that were discussed in 2010 at Harvard's Connecting the Mind-Brain to Education Institute and The UCLA Learning Lab include:

Myth: Basics must be learned so well that they become second nature. **Fact:** Over-learning basics at the start can stifle creativity and individual expression.

Myth: Delaying gratification is important. **Fact:** Keeping on growing interest and joy in learning leads to more learning.

Myth: There are always right and wrong answers. **Fact:** Correctness depends on the context.

Myth: Good learners know what is out there. **Fact:** Life-long learners are not know-it-alls.

Myth: Forgetting is a problem. **Fact:** Memory often prevents new or novel learning and use beyond a personal bias.

Myth: Memorization is necessary. **Fact:** When possible, relating new information to personal experience is better than memorization.

Myth: There is a limit to what can be learned. **Fact:** It appears there is no limit to storing information. Unfortunately, our ability to recall is limited. Therefore, approaches to learning must take care that they support the skill of recalling information. Use metaphors and stories.

Myth: Orderly learning is the aim. **Fact:** Students need unpredictable environments to gain understanding. Struggling creates access to new learning for the long-term. "Order does not establish learning that lasts." (Dr. Robert Bjork)

Myth: We use only 10% of our brain. **Fact:** Most, if not all, of the brain is engaged in learning and everything else we do.

Myth: We are aware of what we are learning. **Fact:** We often are learning without meaning to do so.

Myth: Learning is putting new information into the brain. **Fact:** Learning depends on prior knowledge and assumptions (good and poor) that connect with new data.

Myth: Learning is based on getting it right. **Fact:** Uncertainty, confusion, and struggles support learning.

Myth: Long sessions on one thing are more useful than several short sessions. **Fact:** Several short sessions are more useful than one long session.

Myth: Study or train in one location during one session. **Fact:** Vary locations during a training or studying session.

INCONSISTENCY IS VALUABLE

Human beings use workable and unworkable outcomes as valuable guides for future interactions with the environment. While this sounds counterintuitive, it is a reality. Inconsistent outcomes are important keys to experiencing meaningful learning and developing to our full potential.

When you come across the term "inconsistent" does it give you a positive or negative emotion? Our perception of inconsistency lays at the foundation of experiencing meaningful learning. Before one makes their evaluation of inconsistency as positive or negative, they should keep in mind that nature, in all its wisdom, allows things to exist that help our survival – which inconsistency clearly does. When learning, we need the inconsistency of unworkable outcomes to reach our personal potential.

It is useful to see things not as hard to do, but as things that can have inconsistent outcomes. For example, in my view, golf is not hard, it is inconsistent. Of course, it is important to have a fundamentally sound golf swing. However, no two golf shots are the same; some adjustment (i.e., inconsistency) needs to be made. The value of inconsistencies is lost in a culture that views unworkable outcomes as failures in need of fixing. This view often retards learning and progress.

The view here is that there is no failure, only usable feedback for future reference, based on the nature of inconsistency. By returning to what-to-do ideas (a positive emotion) and avoiding attempts to fix outcomes (a negative emotion), meaningful learning is supported. Unworkable outcomes are "Desirable Difficulties," A term coined by Dr. Robert Bjork that I refer to as Desirable Developmental Difficulties.

> ***In learning situations, elements of inconsistency are tools. If not seen as such, growth, development, and improving can all be damaged. Mistakes, in actuality, serve as information for future reference and adjustments to achieve the desired outcome.***

Student-centered/brain-compatible learning occurs when inconsistent outcomes are seen as essential components of meaningful learning. UCLA's Robert Bjork's paper—*How We Learn and Should Practice, vs How We Think We Learn and Should Practice* provided useful insights about how learning from inconsistencies.

Robert Bjork said, "From a practical standpoint these findings point to reasons instructors are susceptible to using less effective conditions of instruction over more effective conditions (emphasizing poor outcomes). From a theoretical point of view, such findings leave implications for the functional architecture of humans as learners."

Why do unworkable outcomes arrive? Because we are human and, by nature, inconsistent.

Bjork's studies suggest that mistakes and unworkable outcomes should be introduced during training. The golfer Jack Nicklaus has said he deliberately performed mistakes during practice to help learn what to do when playing. Being exposed to unworkable events is more valuable than studying events, as principles outweigh rules.

Learning is a social and emotional process. Student-centered/brain-compatible processes for learning have demonstrated that an effective way to evaluate students, or for students to evaluate themselves, is by moving away from seeing unworkable outcomes as a failure in need of fixing. Unworkable outcomes are often more useful for experiencing meaningful learning than workable outcomes.

UNWORKABLE OUTCOMES

A counterintuitive insight: Learning can be a green-light journey when guided by inconsistency, unworkable outcomes, confusion, and conflicts. These support the choreography of development and curious thinking while struggling to learn workable outcomes in safe, smart, playful environments.

The neurological changes from inconsistencies likely have an adaptive advantage for survival. The goal then is to mute the negative response to failure by redefining unworkable outcomes as an opportunity to learn. Student-centered/brain-compatible educational environments engage the whole human being as the physical, social, cognitive, emotional, and educational needs are being nurtured with a constructively positive approach.

There is a smaller return on the investment of time and resources by trying to fix habits and poor grades, than when we enhance an individual's ability to reach his performance potential. For example, paying attention to, or criticizing, an undesirable outcome is not as useful as helping students learn what workable outcomes are needed; the latter lowers frustration.

Dr. Robert Bjork states, "People learn by making poor outcomes. Introducing mistakes during training is important."

> ***Orderly learning is not the aim, "Students need unpredictable environments in order to gain understanding, thereby creating access to it for the long-term. Order does not establish memories that last." (Dr. Robert Bjork)***

"The consequences of human evolution can be seen in our responses to the inconsistencies in our environment. This adaptive process is at the core of our survival. As environments change, there are only two possibilities for survival: successful mutations propagate, or via rational choices, behavioral modifications increase the odds of survival. An excellent negative example is found in the Viking settlements of Greenland starting in the 10th century. Temperatures were relatively warm; farming was successful. All went well until the Little Ice Age of the 15th century. As temperatures dropped, the Viking settlements collapsed largely because they refused to learn from the longevity of more northern native Greenlanders who had learned how to live in Arctic-like conditions. Adapt or die–there is no other choice!" (Dr. Yazulla)

FAILURE vs FEEDBACK

Failure 101 is an actual freshman course at Penn State in which under Professor Jack V. Matson, the value of failure is explored. Unfortunately, the business of education often backs away from poor outcomes. The view here is that flaws are what are incomplete on their way to full development.

Unfortunately, we are told to avoid failure. We are criticized for failure. We are told, "Don't do that again." We are afraid of failure—why?

One answer is that at some stage of our lives we are programmed to see unworkable outcomes as a failure. This view often prevents instructors and students from experiencing meaningful learning.

Realistically, unworkable outcomes are not an option; they are a fact of life. If you can accept that insight, your future choices are straightforward. First, you can use an approach to learning that does not ignore what can be gained from undesired outcomes, and you can see those outcomes as valuable feedback for future references.

A pie that has been overcooked cannot be fixed, but another pie can be created, based on what was learned from the poor outcome of the first pie. No fixing needed; just develop, create, and invent something different. We think of a poor or unworkable outcome as one that is not able to be used for a particular circumstance. We are now learning that all is not lost from an undesirable outcome as it can lead to looking at the challenge differently or recalling that outcome in the future for an application it might be better suitable to. The crust might not have worked for the pie you were creating, but you might find it is perfect for a different type of pie in the future.

Brain-compatible learning has an appreciation for unworkable outcomes because of their transformative value. When we are learning with the brain in mind, unworkable outcomes have the ability to become emotionally accepted for future reference (non-frustration), not perceived as a failure.

Unfortunately, in some learning environments, students are often told they are doing something wrong that is in need of fixing or that something is missing. In brain-compatible learning environments, students see unworkable outcomes as something that is developing and growing, not as something that is wrong; nothing is missing, it is developing!

Unworkable outcomes are valuable. They are the teacher, or guide, for doing something different, or arriving at a different answer instead of being seen as mistakes.

> ***Unworkable outcomes are a biological requirement of change and are accepted as natural inconsistencies on a positive, proactive journey of learning.***

Efficient approaches to learning are often acting more like editors, taking the best from here or there, and are less about pointing out unworkable outcomes. These proactive, positive paths to change make a journey of learning safe, playful, and appealing, even if they are rough and gritty when learning.

DIFFERENT is the aim, not better. If a foul shot in basketball misses and the next one scores, it was different from the first attempt. The first shot guided the change in the second shot. That is an example of a positive emotional mindset when learning and gaining an education. Trying to fix what is seen as a mistake is an emotionally negative approach that can suppress learning. Just do it differently, accepting and learning from any inconsistencies along the way is a student-centered/brain-compatible approach to change.

Accepting inconsistency is a major emotional tool for learning and developing. A quest of consistency is "fool's gold." It is often said that a team, student, company, or player is good, but to be great, they must be consistent. That is not true. They will out-perform others when they deal with inconsistency as a fact of life; learn, and move on!

Organizations and people reach their full potential not because they are consistent, but because they adapt to inconsistency differently than others.

ERRORS ARE USEFUL

"A man of genius makes no mistakes; his errors are validation and are the portals of discovery." (James Joyce)

The brain does many things well but nothing perfectly. In their book, *Bozo Sapiens: Why to Err is Human*, researchers Michael and Ellen Kaplan offer valuable insights into the nature of assumptions, predictions, doubt, error, and learning.

The Kaplans started with the obvious assertion "to err is human." Over time, humans adapted by testing assumptions and making predictions through trial and error. This process had the brain taking in information from workable and unworkable outcomes, thereby rewiring brain cell connections in an ongoing process.

> ***Assumptions and predictions are more than important thinking tools; they are how healthy brains function.***

Human beings assume and predict their way through their daily lives. We are always harvesting information from our non-conscious minds, which is the underlying source of our learning, developing, and survival skills.

We make predictions and assumptions all the time. We assume something will be good or bad. We assume and predict something will be interesting or not. We assume whether it is the right time to do something or not. We assume what will or will not work, etc.

> ***A healthy brain is efficient. It is filled with general, just in-the-ball-park, detail-free concepts that support learning. Studies show we could not function if we consciously knew all that was going on, hoping to avoid error.***

To learn is to be allowed a choice. To choose includes the option to choose badly. Doubt and error are the starts of wisdom in a world filled with false ideas. To use the Kaplans' words, "to find sense within non-sense, or meaning from meaningless, is the power of error."

Struggle has been found to be one of the main paths to experiencing student-centered/brain-compatible learning and often is not recognized for its value. Having individuals do things incorrectly in learning environments is often the most efficient way to have them learn to create a different or workable outcome.

When we are learning and performing, unworkable outcomes are valuable components of student-centered/brain-compatible learning. Can you uncook an egg? NO! We must start over, doing things differently, based on what was learned from the unworkable outcome of the first attempt. Baseball, golf, and tennis swings cannot be fixed after they are made; only remade differently. The past is irretrievable; move on!

F Finds
A Access
I Into
L Learning
U Uncovering
R Relevant
E Experience

TOP TEN QUOTES ON UNWORKABLE OUTCOMES

The following are Richard Branson's (Virgin Airlines) favorite quotes on the value of unworkable outcomes.

> "Every person, and especially every entrepreneur, should embrace failure with open arms. It is only through failure that we learn. Many of the world's finest minds have learned this the hard way – here are some my favorite quotes on the importance of failure, and the road to success." (Richard Branson of Virgin Airlines)
>
> "When we give ourselves permission to fail, we, at the same time, give ourselves permission to excel." (Eloise Ristad)
>
> "The greatest mistake you can make in life is to be continually fearing you will make one." (Elbert Hubbard)
>
> "Only those who dare to fail greatly can ever achieve greatly." (Robert F. Kennedy)
>
> "Success is stumbling from failure to failure with no loss of enthusiasm." (Winston Churchill)
>
> "One who fears failure limits his activities. Failure is only the opportunity to begin again more intelligently." (Henry Ford)
>
> "The greatest glory in living lies not in never falling, but in rising every time we fall." (Ralph Waldo Emerson)
>
> "Develop success from failures. Discouragement and failure are two of the surest stepping-stones to success." (Dale Carnegie)
>
> "I have missed more than 9000 shots in my career. I have lost almost 300 games. Twenty-six times I have been trusted to take the game-winning shot and missed. I have failed over and over and over again in my life, and that is why I succeed." (Michael Jordan)
>
> "I have not failed. I have just found 10,000 ways that won't work." (Thomas A. Edison)

FLEXIBLE KNOWLEDGE-PORTABLE SKILLS

"How does the mind work and especially how does it learn?" is the question asked by Professor Daniel T. Willingham, at the beginning of his article, "Inflexible Knowledge." (The American

Educator, Spring, 2002) The following was compiled from and influenced by that article.

Mankind's brain has a bias for remembering new concepts and information in terms that are superficial and concrete, not in terms that are abstract and deep; but individuals who have mastered any skills have flexible knowledge, not superficial and portable skills that are not concrete.

An expert is one who, in some endeavor, has achieved a level of competence that greatly exceeds those in the general population. Experts can transport the knowledge and skills they learn in one environment, to dissimilar conditions and contexts. Unfortunately, in some approaches to education people are not gaining the insights and training that allow them to transport what they are learning beyond their current conditions.

Those individuals are learning what science calls, "inflexible knowledge and non-portable skills." They are just being conditioned. On the other hand, in learning environments that are supported by self-development through self-discovery, knowledge becomes flexible and skills become portable.

Approaches to learning that do not give learners the opportunity to use their own curiosity, imagination, and their "what if I did this" ideas, tend to develop information that stays concrete and superficial not deep and flexible.

> ***It is the approach to learning that determines if information becomes flexible or inflexible knowledge.***

The aim of a good education is to enable students to apply new concepts and new information to situations that go beyond classrooms, practice fields, labs, or golf practice ranges. For example, a student with flexible knowledge who wants to paint a narrow line, but is holding a three-inch wide brush, will turn the three-inch paint brush on its side and use the narrow side to achieve his desired results. On the other hand, a student without flexible knowledge will look for a thinner paintbrush. While this is a simple example, it demonstrates the difference between flexible and inflexible knowledge. Development through self-discovery develops the kind of portable skills and flexible knowledge that following directions does not develop.

"Memorizing with an absence of meaning," would be an informal definition of rote learning, which is the bogeyman of education.

"We would like students to gain insights and meanings. A good education seeks to develop creative problem solvers, not parrots. In so far as we can, we must prevent students from absorbing information in the rote form." (Using drills and copying expert models are examples of rote learning found in math, history, grammar, sports, etc.) (Professor Daniel T. Willingham)

Of course, mastery of basics is critical for a firm foundation. However, it is the flexible application of these basics that make up a meaningful education.

Some instructional methods condition students with rote learning techniques. There is a difference between learning and conditioning. When we are encoding and learning, we are not just gathering rote information without insights and meanings.

Conditioned students have narrow, inflexible knowledge, lacking any deep structure that is portable. For example, many golfers who lack flexible knowledge and portable skills overlook the fact that the principles for applying and aligning force in the putting motion are similar to the motion of a longer swing.

A self-discovery approach to learning gives people the opportunity to become their own best teachers by enhancing their ability to respond playfully to ever-changing environments as they expand and construct flexible knowledge and portable skills.

Webster defines learning as, "acquiring knowledge by experience," and education as "training by which people learn to develop and use their own mental and physical skills." The joy of learning and gaining a good education does not come directly from teachers; it is based on nature's plan for self-development through self-discovery.

Traditional memorization approaches to instruction can limit possibilities. On the other hand, nature's plans for efficient learning evoke unlimited support for the arrival of new insights.

Instruction is often over-dressed with steps that do not create a strong link between a learner and his environment. Traditional instruction can be like standing on a slippery slope and trying to get to an exact point, instead of being fascinated by the journey. In learn-and-develop environments, students are softly looking for all the possibilities; they are not looking hard for an answer.

Suggestion! *Be a private investigator and report to yourself what you notice about your environment. Become your own reporter, not a follower, who looks for someone else's impressions. Good coaching, mentoring, or guiding will help people notice what is being overlooked. Everything we notice has meaning and input for future reference.*

Everything has application depending on your point of view. What your brain is encoding is either workable or unworkable; both are valuable for learning. Knowing the left and knowing the right helps to identify the center.

> ***There is no failure; only accurate feedback in learning, developing, and growth environments.***

Harvard's Lawrence Hooper said, "The ultimate source of exceptional performance is exceptional learning." I always mention learning first, because how we learn is the turning point for how we should approach teaching, which then influences performance. We are wired to learn first, then teach ourselves to use what we have learned. For example, after we learn about something, we teach ourselves to avoid it or put it to use, and that process is influenced by the approach to learning being used.

RECAP

Before adhering to a theory, be sure it has cross-research behind it.

Teach for an outcome of what you want the receiver to learn.

Mastery of basics is critical for a firm foundation, BUT meaningful education is derived from the flexible application of the basics.

A self-discovery approach to learning gives people the opportunity to become their own best teachers by enhancing their ability to respond playfully to ever-changing environments as they expand and construct flexible knowledge and portable skills.

There is no failure, only usable feedback for future reference, based on the nature of inconsistency. Organizations and people reach their full potential not because they are consistent, but because they adapt to inconsistency differently than others.

YOUR NOTES ____

____

____

CHAPTER THREE

Views on Learning and Teaching

"Mainstream education around the world demonstrates a disconnect between the way humans learn and the way they are taught." Sarah Weiss (Brain World, Spring, 2015)

"So few educators have had formal development in teaching practice." Studies show that there would be less talk about poor schools, poor grades, poor sports skills, and poor performing if both teachers and students put more attention on the process of learning than on trying to fix unworkable outcomes. (Professor Terry Doyle)

"Learn in harmony with your brain," is the suggestion Doyle makes. "There is nothing in our life, workable or unworkable, expected or a surprise that is not organized and influenced by the brain." THERE IS ALWAYS A BRAIN CHANGE BEFORE A PERFORMANCE CHANGE. (Professor Terry Doyle)

Allen Watts, the author of The Book, asked, "What do we have to know to be in the know?" I ask, "What should instructors and students know to support efficient learning in any learning environment?

CULTURES AND CUSTOMS INFLUENCE LEARNING

To clarify the statements in this segment, the terminology refers to definitions as follows:

CULTURE: As related to the field of Education. Long held elements believed to be necessary.

CUSTOM(S): Traditions accepted and passed along within a given household, community, etc.

ENVIRONMENT: The surroundings or conditions in which a person is raised. Organized or disorganized? Functional or not? Emotionally and physically safe or not? Supportive or not?

Customs and cultural views can change quickly, especially when it comes to ideas about education and learning. Over time, views about education have changed many times, often quickly in the span of a few years.

To reach our potential as instructors and students, customs and cultures cannot split life into living and learning. The methods used by approaches to learning should incorporate data obtained from wide sources of research including Social, Behavioral, Cognitive, Biochemical, Evolutionary, Psychology, Psychological, and the Neurosciences. These fields of study have resulted in data for meaningful models to optimize learning and performing.

The culture and customs of student-centered/brain-compatible approaches to learning carry a message that supports learning while other approaches have been found to put forward a message that can be counterproductive, suppressing long-term recall of information and skills.

There is a smaller return on the investment of time and resources from a culture that is trying to change unworkable outcomes (i.e. poor habits, poor grades), then when we enhance an individual's ability to reach their performance potential by learning from our unworkable outcomes.

The culture of any approach to learning influences the neural pathways that guide the processing and comprehension of information. The culture and the environment that a student is raised in influence whether the person becomes a dependent or an independent learner. Independent learners develop their own intellectual capacity through curiosity and the intrinsic motivation to satisfy that curiosity, perhaps under the guidance of supportive instructor(s).

Unfortunately, dependent learners rely on instructors to carry the responsibility of doing most of the thinking during learning opportunities. This is often referred to as "spoon-feeding," in which students are told exactly what they need to know, without critical thinking. This is not an approach that is compatible with stimulating innovation and self-determination.

"It is our education which makes the great difference in mankind." (John Locke; 1632-1704)

Many of John Locke's views on education were compatible with the brain's connection to learning. He did not see much value in what he called "bookish" learning. Locke's aim was to develop students that were thinkers, and not just gatherers of information. Locke was turning teaching environments into student-centered/brain-compatible learning environments. What John Dewey called "democracy in education," is also student-centered/brain-compatible for learning.

J. Gilbert, Google's Director of Talent, once said, "Problems can never be understood or solved in the context of one academic discipline. Approaches to learning should be interdisciplinary." Learning is a social, cultural event; it alters behavior; grows the brain; is a chemical process; an evolutionary journey; influenced by emotions.

Culturally and individually, we have been seduced by the perceived value of intellectually interesting information and by teaching-fixing to get something right methods of teaching. Both side-step brain-friendly, student-centered, learning, developing, and growth approaches.

SECTIONS OF THE BRAIN

Different sections of the brain do different things. Some parts help you learn and remember; other parts help you to solve problems and make decisions. Some help you with balance or to move your hands, arms, and legs. Your brain does all these things and more. Your brain is amazing; to fuel its development requires 43 percent of the body's daily energy intake (calories) until the early teens.

"Students do not have a blank slate for a brain; it is a customized organ encoded with their experiences which influence learning more than the information to be learned. Even the youngest learner's brain has been shaped and wired by home environments, siblings, extended family, playmates, genes, trauma, stress, injuries, violence, cultural rituals, expectations, enrichment opportunities, primary attachments, diet and lifestyle." (Eric Jensen) Efficient approaches to learning keep this insight in mind.

People make problems of unwanted outcomes by interpreting them as failure instead of seeing them as just a condition or a location on a journey of development. Learning, like art, is a creative process that flows from one step to the next as it evolves.

> ***Each instructor and student is unique with a different background. This requires taking the information here about doing learning and personalizing it, depending on the circumstances and environments.***

There are times when learning takes hold and times when it does not. The suggestion here is to become informed about the brain's connection to learning and make that a "turning point" in the direction of designing information-delivery systems that are compatible with the nature of learning.

It is every reader's choice how to best apply studies about brain-based learning approaches. A reliable self-discovery (trial-and-error) approach becomes brain-compatible when poor outcomes are not criticized. This non-critical approach was used by Peter Drucker, a noted educator, who gave test papers back to his students until they warranted a grade of an "A." "This progress has a greater emphasis on recognizing good behavior than on punishing bad, thus lowering frustration. (Dr. Raver, Psychologist, New York University's Chicago School of Readiness Project)

PROGRESS

It is surprising what can be misunderstood about our natural ability to learn when the culture sees teachers, coaches, parents, or business training programs as the primary source of information and answers for the problem to be solved without taking the brain's connection to learning into consideration. If approaches to education are going to improve, you cannot stand outside and watch and expect to understand learning. Suggestion: look at and become informed about the nature of learning and the brain's connection to that process of change.

A student centered/brain-compatible process in the direction of meaningful learning is emotionally captivating and more a function of full awareness than direct instruction.

It has been found that some approaches to learning offer up little more than a proxy of learning, side-stepping real contact with a student's potential for gaining the kind of wisdom found in "flexible knowledge and portable skills." (Professor Daniel Willingham, Psychology, University of Virginia)

Meaningful learning is influenced by more than meets the eye; it is an internal process that guides an external outcome after an internal change. What some cultures do not recognize is that what is going on inside the brain matters, and this should be taken into consideration when learning, developing, and performing. Things we do mentally and physically change neuron connections inside the brain.

Gaining an education should be accomplished in a way that maximizes the communication of information and minimizes the complexity of learning. Said another way, the culture, context, and environment that approaches to learning and human development take place in should be free of frustration and intimidation.

What is needed is a culture that recognizes the value of learning and of acquiring know-how skills, not just information. Knowledge is applying information appropriately to a contextual situation. On the other hand, it can be a way of expressing information such as numbers, charts, and statistics that have little value in and of themselves. The nature of learning and the brain's connection to change normally do not require data-driven instruction.

WHAT I LEARNED

From neuroscience:
The nature of learning is seamless. It touches the physical, mental, and emotional components of learning all at the same time. Meaningful learning takes place at the intersections of before, during, and after acts of learning, developing and performing.

From Dr. Kurt Fischer of Harvard's Graduate School of Education:
Accurate subject content information is not the challenge. How approaches to learning are designed and organized to deliver information is the challenge.

From Eric Jensen's "Teaching with The Brain in Mind Workshops":
The culture and customs of efficient approaches to learning are teaching with the brain in mind.

From cognitive science:
Our capacity to recall information is much smaller than our brain's seemingly limitless capacity to encode information. Therefore, it is important that approaches to learning enhance our ability to understand and recall information. Memory is more an act of rebuilding than recalling.

From Dr. Bjork:
What takes place during lessons, workable or unworkable, is not an indication that learning has or has not taken hold.

From developmental research:
Follow the bouncing ball from information in the environment to the brain, to prior knowledge, to interactions, to workable and unworkable outcomes.

Student-centered/brain-compatible approaches to learning support meaningful learning; they do not guarantee learning.

From studies into emotions:
Doing things for their own sake, not for a reward, emotionally supports the nature of learning and curiosity about information. The power of the limbic brain (emotions) is far reaching and is always a large component of learning.

Both appropriately and poorly managed information and skills start with how the approach to learning interacts with the internal workings of the brain.

All outcomes in learning environments are influenced by decisions made at the beginning of the process. Any information being shared and all outcomes that are set into action begin with a decision based on assumptions that instructors and students have about learning.

> It took me a few years to recognize the value of golfer Fred Shoemaker's statement, ***"what if we never fixed anything."***

A BRIDGE

"Form follows function" is a universally accepted principle. The function of approaches to learning is to give students the best opportunity (form) to gain and experience meaningful long-term learning, including developing know-how skills.

Keeping the brain in mind in learning environments is the bridge to the principle of "form follows function." My personal experience was that I overlooked this insight about the brain for my first 20 years of teaching—teaching, but not with the brain in mind.

"Learning is an inherited mechanism that serves to promote the survival of our species. We are mostly learning by nature and not by reason and without conscious thought. Learning is a reflex, an involuntary motivational form of survival. It is an inborn pattern of behavior that is responding to environmental stimuli. Learning is an innate capability impelled from within." (Konrad Lorenz and Nino Tinbergen)

The nature of learning is a sea-worthy vessel that is floating away from intimidation and frustration in the direction of learning from prior knowledge. Prior knowledge encoded in the brain could be seen performing like a reservoir of moving, ever-changing information, not a library filled with inactive information.

This reservoir holds an astonishing and diverse volume of unusual conditions that were once seen as problems but were gifts filled with messages that support the development of new learning.

When memory is referred to as a library where thoughts and ideas are stored for future use, I see a static collection of information. But in a reservoir-like memory, what is stored there is always in motion, reacting to ever-changing real-world conditions. Different thoughts and ideas come together, cross-fertilize, and come into play before external outcomes, both workable and unworkable can arrive.

Learning should not be seen as an event. Learning is a process that develops over time. Biological possibilities are revealed during a playful journey of discovery and not in the observable results or outcomes of our interactions with our environment. A process of personal discovery leads the way to outcomes, gaining insights into the process that we support when experiencing workable outcomes.

FIRST THINGS FIRST

"First things first," is a popular statement. In any learning environment, having accurate insights into the nature of learning and the process of change are important steps for educators, students, employers, parents, and coaches of any topic.

Learning is about experiences and the outcomes those experiences produce. Unconscious, implicit learning, or curious active self-learning, is our main form of meaningful learning. On the other hand, explicit approaches to learning (information passively gained from others) can be the poorest form of learning.

If Meaningful Learning is to be experienced, just sharing accurate information about subject content will be much too limiting. Subject information must be moved in and out of a variety of contexts in a way that connects with similar and dissimilar prior knowledge. That way learners develop information and skills for use in different environments. This approach takes the brain's internal connection to learning into consideration whereas just sharing information about subject content does not.

> ***Mihaly Csiksentmihaly, the author of Flow, would ask: Was the approach to learning enjoyable, allowing students to exercise a sense of control over their actions?***

In his book, *How to Fly a Horse*, (p. 38) Kevin Ashton had a wonderful description of thoughts. "Thoughts and ideas start as caterpillars of the conscious mind, become cocoons in the non-conscious, and then fly out like butterflies. The key to creating and learning is to cultivate more of these moments."

"One hundred thousand years ago, at least six different *Homo* species inhabited the earth. Today, there is just one, us, *Homo sapiens*; how did our species succeed?" Yuval Noah Harari, the author of *Sapiens, A Brief History of Humankind*, provided an answer, "Human beings are a brain-based species, wired for survival and the ability to learn, not fail."

Student-centered/brain-compatible information-delivery systems support a student's ability to learn playfully, both now and in the future. These approaches to learning create a team made up of facilitators (teachers) and participants (students).

Student-centered/brain-compatible strategies are designed so that acts of learning become playful vehicles for applying the nature of learning. "These acts include the steps and stages of analyzing, drawing conclusions, posing questions, collecting data, reasoning, and deduction skills." (Archibald D. Hart, *Habits of the Mind*; 5)

RECAP

Customs and cultural views and ideas about education and learning have changed many times over time.

It has been found that methods used by approaches that expect students to reach their full potential to learn should incorporate data obtained from wide sources of research including Social, Behavioral, Cognitive, Biochemical, Evolutionary Psychology, Psychological, and the Neurosciences.

The culture, context, and environment that approaches to learning and human development take place in should be free of frustration and/or intimidation.

A culture that recognizes the value of knowledge-driven learning in the direction of know-how skills supports the student's ability to experience meaningful learning.

Meaningful learning is applying information appropriately to a contextual situation.

Learning is gaining experiences from the outcomes those experiences produce.

Subject content must be moved in and out of a variety of contexts in ways that connect similar with dissimilar prior knowledge, thereby developing information and skills for use in different environments.

YOUR NOTES ____

____

____

CHAPTER FOUR

Early Educational Research

"All who have meditated on the art of governing mankind have been convinced that the fate of empires depends on the education of youth." (Aristotle – almost 2400 years ago)

"The history of education affords a perfect key to the history of the human; the line of human progress. A line from its humble beginning to the precious heritage of the present." (E. Jensen, Brain Based Learning; 7)

This chapter on the evolution of Educational Research has two aims. First, to share the volume of studies relating to education with readers who may have been unfamiliar with those efforts. Next, as this area of research expands it hopefully will continue to become available and put into practice, thereby enhancing learning opportunities.

LOOKING BACK

The oldest existing, and continually operating educational institution in the world is the University of Karueein, founded in 859 AD in Fez, Morocco. The University of Bologna, Italy, was founded in 1088 and is the oldest one in Europe. Today, it is a public university. (Guinness World Records)

Harvard University, founded in 1636, claims itself to be "the oldest institution of higher education in the United States." The claim of being "the first university" has been made on its behalf by others. (brainly.com)

The College of William & Mary in Williamsburg, Virginia was founded by royal charter in 1693, making it the second oldest college or university in the United States.

The Learning & Developing Book, by Tricia Emerson and Mary Stewart, refers to the history of educational research as having a beginning, middle, and end.

> **Beginning:** (1896-1980) ideas were often based on research by the famous Jean Piaget, whose 20th century studies of his own children to determine how the brain stores information led to his four stages of cognitive development.
>
> **Middle:** (1940-1960) ideas were developed by David Ausubel in 1960 as frameworks for processing and storing new information.
>
> **End:** ideas were being validated by researchers as an effective way to improve understanding and recall. Hopefully, there is no end to the development of effective, student-centered/brain-compatible communication.

Humans communicate accumulated knowledge through art and by oral tradition (story, song, lecture) and in writing, either formally or informally. The method, culture, and environment in which information is transmitted greatly affect learning. Educational institutions have grappled over the most efficient methods for centuries.

In her must-read book, *Building A+ Better,* Elizabeth Green discussed the early history of educational research. Today, in some environments, 21st century neuroscience has had a large influence on designing approaches to learning. However, how did we arrive where we are today in the field of education research?

What follows is based on Green's findings.

Prior to 1940, serious studies on teaching did not exist. In 1948, Nate Gage landed his first academic job at the University of Illinois in the field of educational psychology. Professor of

Education was his title, but he did not take to teaching. Instead, he turned his interests to doing research on the topic of teaching.

Professor Gage and other researchers at that time suggested that good teachers should be "friendly, cheerful, sympathetic, and morally virtuous rather than cruel and unsympathetic. However, they were not yet looking into acts of teaching.

In America, by the 1950s, there were university programs that trained and certified individuals as teachers in what were called "education schools." However, these students of education were not learning anything about teaching. How was this happening? Education schools were neglecting the heart of the profession, acts of teaching and learning.

Green points out that one of the reasons early education training ignored teaching methodologies was that it was said to be uninteresting. The early research looked into hundreds of variables but ignored the teaching process. Nate Gage concluded that studies would only unlock the mysteries of the American classroom by going inside of them. He believed that he was going in the right direction to construct a true science of learning with that insight.

Gate called his method the "process-product paradigm." By comparing teaching to learning results, you were comparing a process to product. Researchers could conclude which teaching acts were effective and which were not. Today we know this is not true. This method caught on quickly and was used for years. Then in the 1970s, Lee Shulman, a researcher at Michigan State, said Gage's method was the same old "B.S."

When Shulman was asked why it was "B.S.," by the National Institute of Education, he answered that this method is a testimony to the past and that "behaviorism" was ignoring the mind. Learning was not meant to be responding to repeated rewards and punishment. Researchers now realized that to explain how people learn; they had to reckon with cognition and the brain.

Shulman, along with others in the field of education found the topic of thinking fascinating. He was now paying attention to the mental operations that take place when a person moves from impressions, to questions, to understanding. By gaining insights into thinking or the process of making knowledge, scientists would no longer just study schools; they would help improve them.

> ***Teachers and students were now looked upon as "information processors." To study teaching by learning about thinking started a journey in the direction of insights into our brain, the gateway to learning. By studying the process of teaching, Gage, Shulman, and others were about to expose many poor ideas about how teaching and learning worked.***

By the 1970s it was being recognized that educators not only had to think; they had to think about how students were thinking. Educators were becoming aware that they not only had to consider what it meant for them to know something, but they also had to find ways to efficiently transfer their knowledge to their students. Research suggested that teaching had to be more than just providing information. As Green points out in her book, teaching was becoming the science of all sciences, the art of all arts.

A JOURNEY LED BY CURIOSITY

Spending time in bookstores, especially ones filled with rare old books is something enjoyable as I am always looking for a book or two to add to my library. Several years ago, I found myself in a rare book store in London curiously doing my best to find some interesting reading about learning theory and practice. Looking around, *The Common School Journal*, published in 1839, caught my eye.

Published in Boston by Marsh, Capen, Lyon, and Webb, this two-volume set was edited by Horace Mann, then the Secretary of the Massachusetts Board of Education. The books contained papers published on the topic of improving education in the Boston schools from over 175 years ago.

The book's introductions are called The PROSPECTUS, and Volume I started with "These papers to be devoted to the cause of Education."

Other information on the first page of The PROSPECTUS included: "Twenty-four is the number of the papers issued each year, sixteen pages each, making an annual volume of 384 pages. The subscription price will be One Dollar a year. The Paper will explain, and as far as possible, to enforce upon all parents, guardians, teachers, and school officers, their respective duties towards the rising generation." What a great statement!

"The Paper will be kept entirely aloof from partisanship in politics. It will address physical health and cultivate to strengthen intellectual faculties. It is not so much the object of this work to discover, as to diffuse knowledge. In this age and country, few things on the subject of education are known."

Some of these statements about education made in the mid-1800s often hold true today. Fortunately, because of the Learning Sciences founded in 1990, we now have information for improving education that was not available to the curious and concerned authors of *The Common School Journal* in 1839. For example, making approaches to learning more in step with learning information and skills for everyday use is not a new view.

In the 19th and 20th centuries, we did not know a great deal about how people actually learn. Approaches to learning were designed based on assumptions that had never been scientifically evaluated. These traditional approaches are known as "instructionism." Beginning in the 1970s, research showed that most approaches to learning and teaching were deeply flawed.

Over the next 20 years, scientists who studied learning began to realize that they needed to develop new scientific approaches that would go beyond what each individual science was capable of uncovering. They began to collaborate with others fields of science.

During "The Decade of the Brain" (the 1990s) and subsequent years, the Neurosciences started to uncover more about the brain than at any other time in the history of research. In the 21st century, research centers around the world are involved in what is referred to as the "Human Brain Project" in an effort to expand insights about the functional areas of the brain.

The Learning Sciences were born in 1991 when its first international conference was held, and *The Journal of Learning Sciences* was published. The Learning Sciences are made up of cognitive science, educational psychology, computer science, anthropology, neuroscience, biology, sociology, and other fields of science. The Learning Sciences are an interdisciplinary field that studies teaching and learning with the goal of enhancing learning environments, to allow people to learn more quickly, deeply, and efficiently.

Learning scientists reached a consensus on some basic facts about learning and teaching as reported in the *Cambridge Handbook of the Learning Sciences*, edited by Keith Sawyer (2006).

This 600-page resource has contributions and research from sixty-two leading universities including: MIT, Harvard, Washington University, Georgia Tech, Northwestern, Michigan, Carnegie Mellon, Vanderbilt, California, Pittsburg, Drexel, Hawaii, North Texas, Stanford, Southern Illinois, Indiana, Utrecht, and several research centers including: Ontario Institute for Studies in Education, Adobe Systems for Technology in Learning, and the Institute of Education, London.

Between 2003 and 2006, the National Science Foundation funded nearly $100 million in grants to accelerate the development of The Learning Sciences. Today more and more individuals in and out of education realize that the research emerging from The Learning Sciences has great potential for improving instruction and learning. For example, research has discovered that computer technology only benefits acts of learning when used with what is known about how we learn.

> ***The Learning Sciences now has over 40,000 members. The United States National Institutes of Health spends $4.5 billion annually on neuroscience research plus an allocation of $100 million for new research to map the brain in 2013.***

WHAT'S IMPORTANT?

The topic of learning is important and far reaching. Studies at Columbia University Teacher's College and Harvard University's Graduate School of Education have shown that the design and structure of the approach to learning that individuals experience influences their future employment opportunities, living conditions, physical and mental well-being, and their pace of progress in sports and other endeavors.

For years, many leading researchers concentrated more on how the brain organized itself over time than on how the brain operates when we think and learn. How do we optimize learning opportunities? —by optimizing the process of learning.

How do we optimize the process of learning? —by recognizing the brain's connection to the process of change and our predisposed ability to learn. We might not have ever seen ourselves as perfect teachers, but as human beings and perfect learners!

There are many educators with a long history of making efforts to enhance education. They have seen old ideas reworked and new views put into play, often with little recognizable progress. There has always been an ongoing effort to improve student learning.

New strategies, suggestions, methods, and theories for improving approaches to learning have come and gone over time with some good to be found in some of them. "Battles between different philosophical camps in education are nothing new whether it is knowledge vs skills, memorization vs project-based learning, or small schools vs comprehensive ones." (Education Week, 12/10/14)

It is fair to say that, at their foundation, the above strategies did not have the advantage of using the insights from research about how the brain learns. When ideas about learning were being developed, research from neuroscience about the nature of learning may not have been readily available. Discussed here are the kind of studies that did not start to become available before the late 1980s.

RECAP

In the past, it was not the brain; it was the teachers and students who were looked upon as "information processors."

In studying the process of teaching, Gage, Shulman, and others were about to expose many poor ideas about how teaching and learning actually worked.

Studying about thinking and learning started a journey in the direction of insights into our brain, the gateway to learning.

The Learning Sciences (cognitive science, educational psychology, computer science, anthropology, neuroscience, biology, sociology, and other fields of science), founded in 1991, is an interdisciplinary field that studies teaching and learning with the goal of enhancing learning environments, to allow people to learn more quickly, deeply, and efficiently.

The educator's responsibility is to use appropriate methods for sharing accurate information thus supporting and enabling the student's goal to actively participate in learning.

It is the responsibility of the Institutions offering degrees in Teaching to incorporate a complete and a contemporary introduction to the Learning Sciences.

YOUR NOTES ____

____

____

CHAPTER FIVE

The Nature of Learning

Development, growth, and adaptation to changing conditions are essential for the survival of all organisms. Humans, however, differ in that what we have learned can be transmitted across generations by symbolic means, thereby sparing our descendants from re-inventing the wheel.

"The great aim of education is not knowledge, but action." (Herbert Spencer; 1820 - 1903) Are the actions of approaches to learning keeping students eager, engaged, active learners? Students will want to learn and be involved when approaches to learning use a process that is playful, emotionally, and physically safe, and compatible with the brain's connection to learning.

NATURAL LEARNERS

It is estimated that all humans, chimpanzees, and the other great apes are directly descended from what occurred in evolution between 5 and 7 million years ago. The ancestor of modern humans, *Homo sapiens*, lived between 50,000 and 100,000 years ago. Thanks to what our ancestors experienced, and the DNA they passed on during evolution, we all come into the world as predisposed natural learners. Learning is a natural survival skill. Fortunately, we were all born to do learning efficiently.

Over millions of years, the nature of learning has moved from just doing acts of trial-and-error survival onto developing skills that today include complex thinking, influenced by our emotions, and experiences. After leaving the safety of caves, look how far we have come—splitting atoms, unraveling DNA, creating robots and landing on other planets! Again, that humans are programmed to learn efficiently is an insight that is often overlooked or suppressed by some approaches to learning.

Daniel Coyle, author of *The Talent Code*, points out that right now teachers, parents, and coaches in our society tend to focus their attention on teaching the material—whether it is algebra, soccer, or music. This is the equivalent of trying to train athletes without informing them that muscles exist! It is like teaching nutrition without mentioning vegetables or vitamins. You may eat properly but have no idea why. We feverishly cram our classrooms with whiz-bang technology but fail to teach children how our own human circuits are built to operate.

It is all completely understandable, of course. Our parenting and teaching practices evolved in an industrial age when we presumed potential was innate; brains were fixed (just as we presumed smoking was healthy, and three-martini lunches were normal); but that does not make it right. In fact, some would argue that teaching a child how their brain works is not just an educational strategy—it is closer to a human right.

THE NATURE OF LEARNING

There are plans for producing products, constructing buildings, and manufacturing other items. These plans are guided by the desired end product. When meaningful learning is the outcome you want, the process, not a plan should take into consideration the brain's internal interactions with incoming information.

We could say that we have a calculating brain and an adventurous mind. The calculating brain is a reliable consumer of information. The adventurous mind can be less reliable but is ideal for developing and performing because it is curious and intuitive.

"Our desire is to incorporate the new discoveries about how the human brain learns, teaching in harmony with the human brain." (Professor Terry Doyle)

Here are some short student-centered/ brain-compatible insights about the nature of learning.

The nature of learning looks for a balance between what can work and what can be understood. Learning will not arrive before students understand the information that is being shared.

The nature of learning realizes that the mind-brain connection seems to learn best from metaphors and stories that bring prior knowledge into play, often unconsciously.

The nature of learning is like museums where you are guided by your own curiosity to have interesting interactions. Interactions cause active access to new insights and development.

The nature of learning expresses the energy that supports inspiration filled with proactive, provocative, personal, penetrating, internal, interpretive, interesting, and interactive elements.

The nature of learning supports self-belief regardless of the content of the task and sees individuals in terms of future gains. It supports participation over perfection and acceptance of unworkable outcomes over criticism and judgment.

The nature of learning wants you to wonder. Learning is encoded long term when the journey is a curious experience; not one that is just following directions. There is no real access to curious thoughts; they just arrive naturally on an as needed basis.

The nature of learning creates things that have repeating value. It realizes that unworkable outcomes are valuable tools that can be managed without much, if any, help from conscious thoughts.

The nature of learning recognizes that when conditions for development exist, neither direct conscious thinking nor external aids are needed and, perhaps, they are harmful to learning.

The nature of learning recognizes that when a thought shows up, most of the time it was smuggled in without conscious effort. When emotionally safe conditions for development exist, direct aid often is not required and it may be more damaging than just a little guidance that does not provide answers.

The nature of learning improves things because of what some want to incorrectly label as a failure. These failures are simply unworkable outcomes that serve as valuable feedback for future reference. They are productive. "Productive failure is an essential part of learning." (Kapur, *Education Leadership 2008,* 11/2014; 20)

The nature of learning has a culture that supports the idea that meaningful learning is a transformative adjustment. In this culture, all accomplishments arrive as if they were always there, waiting to be uncovered and drawn out from the indirect preparation that prior knowledge provides for future learning.

The nature of learning recognizes that where we have been in the past, determines if we will travel forward, backward, or nowhere at all, depending on the story we tell our self.

The nature of learning recognizes that our mind-brain activity and trial-and-error adjustments are partners. Being ready and being prepared are not the same. Many individuals say they are ready, but less are prepared. Workable outcomes come from awareness, not from trying. Seeking perfection often gets in the way of workable outcomes.

The nature of learning recognizes that "something old, something new" is a statement which can be used in many areas of life, but to the nature of learning, this is where everything starts. Meaningful learning is a process of old information evaluating new information first before it is either rejected or welcomed into our memory.

The nature of learning recognizes that fear and doubt have three times the influence on learning than the positive effect of joy.

The nature of learning wants you to gather, develop, and invent your own proactive impressions and rethink current negative ones. Things can be seen in a positive or negative light.

> ***The nature of learning recognizes that all events are neutral; it is your reaction to them that gives them their emotional power. Life is 20% what happens to you, and 80% how you react to it.***

The nature of learning is like a small town where everybody talks to each other at the same time. A crowded, chatty community center is similar to how brain-cell connections simultaneously share information with each other.

> ***The nature of learning can be mobile and disorderly with lots of contrasts that inspire creativity. More often than not, meaningful learning is built on an economy of words and thoughts that get right to the essence of things.***

The nature of learning recognizes that when you buy in bulk it can seem like a good deal, but often when it comes to learning, information in bulk is not a bargain or that good of a deal. The Big Gulp®, the Double Whopper®, the Triple Burger®, in all their bloated portion sizes have excessive calories; how many of them can we take in and still stay healthy? Information can be emotionally seductive and intellectually interesting, but how much is useful for experiencing meaningful learning? Often less than is available.

If the nature of learning could speak we might hear, "Our eyesight may not be that important, it is our perception of what we see that is important." We create our own meaning, a private framework that influences learning. Learning starts with either what is happening, or from information and our feelings about the event. The words, thoughts, and behavior that arrive during our perception are always influenced by our physical and social surroundings.

The nature of learning recognizes that people will stop themselves internally, long before they do externally. Gaining an education is about what happens from the neck up that enhances what can happen from the neck down.

The nature of learning knows that learning is a process of becoming that makes it possible for students to become themselves. Just a little attention to the process can make a major difference in the quality of learning.

The nature of learning wants strategies for learning to build confidence in student's ideas while they explore different perspectives. This means embedding in approaches to learning frequent opportunities for students to think and problem solve. Thus, rather than giving students information, instructors should question, prompt, and cue their thinking.

> ***Suggestion!*** *Don't change your vision; change your way of getting there. How we ask ourselves questions leads to the answers we get. Trying to be fully educated often gets in the way of becoming educated. When learning, you are arranging for what lies ahead.*

Meaningful learning comes out of us, not into us. When learning is approached with the brain in mind, it is "out-struction," a term teachers Pia Nilsson and Lynn Marriott used years ago. Learning, development, and growth come out of who we are from moment to moment.

"We do not come into this world; we come out of it, just as leaves from a tree. As the ocean waves and the universe peoples. Every individual is an expression of the whole realm of nature, a unique action of the total universe." (Allen Watts, *The Book*)

Life of the Mind, by Hannah Arendt, "Thinking–a process we take for so obvious and granted, yet burdened with complexities and paradoxes that often keep us from seeing the true nature of reality." Meaningful learning, more a product of wide observation, than the result of directly focused thinking.

THINK FOR YOURSELF

"The tools of contemporary neuroscience are beginning to identify the brain's mechanisms that govern learning, memory, and success. These brain mechanisms help individuals gain the kind of education that can promote learning more than is available from just books and lectures.

This includes the advancement of reasoning and deduction skills." (Stephen M. Fleming, cognitive neuroscientist based at New York University and University of Oxford)

Many traditional approaches to learning have misled students by doing most of the thinking for them. After depriving students of fruitful opportunities to use their brains, we complain that they do not know how to think or problem solve. The aim, of course, is to help students to become independent learners one day. Telling students how to do things not only hurts their intelligence, it can also create confusion. Teachers should help students learn how to learn, and not give them answers.

For example:

> In learning environments did information come from giving individuals choices or giving answers and directions?
>
> Did the information arrive during a random ever-changing approach to learning or during a structured detailed approach?
>
> Was the information just beyond the current skill level of the receiver of information?
>
> Did the information come from what-to-do information or how-to directions?
>
> Did the information arrive from changing a student's insights, or from trying to change outcomes?
>
> In what context was the information provided, and did it come from criticism, judgment, or corrections?
>
> Did the information come from being aware of the process of learning, or from focusing on a result?
>
> Was the information provided through metaphor, or from a list of details?
>
> Was the information gained through self-discovery, or was it provided by an expert?
>
> Was the information uncovered and invented, or was it given to students?

TEAM EFFORT

In the music industry, a group can be described as "bandmates," in sports, it is "teammates," in business it is "staff." When it comes to experiencing meaningful learning, there is also a group effort with the team led by the brain. If the brain could talk it may point out, "When discussing learning, include the elements of memory, emotions, organization, context, environment, experiences, methods being used, and the brain. Learning is a group or team effort."

Memory includes both internal and external conscious effort as well as the non-conscious "aha!" insight. Memories are built, not recalled. Memory is what our ancestors used to be able to communicate tribal knowledge before writing was invented.

Emotions: learning is an emotional event. How individuals feel about what they are learning; how they are being asked to learn; where and from whom they are learning from matters.

Organization: there is a human need or biological imperative to organize our lives, thoughts, and environments, which in turn supports new learning. The brain is flexible and organizes itself. Evolution does not design things or build systems; it settles on naturally organized systems that have a survival benefit." (Daniel J. Levitin, *The Organized Mind*; xix)

Context: learning that takes place in a similar context to where it will be used is normally more efficient than when it does not. Many so-called learning aids cause cognitive blocks to occur because they are not in context and are less valuable for new learning than may be believed because they eliminate struggling.

Environment: learning and developing are supported when the environment in which learning takes place is emotionally and physically safe.

Experience: all new learning is influenced by what has been experienced with both workable and unworkable results. New information is evaluated both consciously and non-consciously relative to experiences determining the new input as either useful or not before it is passed on to long-term memory. All events in our life become indirect preparation for new learning in the future.

Approaches: information that comes into the brain is decoded when learning. Only the information that makes sense and has meaning to a receiver moves on to long-term learning, waiting to be recalled. Unfortunately, some approaches to learning do not take this process into consideration.

Brain: it is the gateway to learning and everything else we do. "The brain is taking in inputs from the world and transforms them into models of the world. The brain is an information processor." (*The Future of the Brain*, by Gary Marcus and Jeremy Freeman)

"GAPS" is a term Eric Jensen used in several of his "Teaching with the Brain in Mind" Conferences. It could be interpreted as:

GAPS = Gone Are Possible Solutions

'thy human beings are natural learners. Unfortunately, because of misconceptions and myths about approaches to learning, GAPS are often created between a student's natural skills as a learner and his level of achievement. The mindset of instructors and students either creates or side-steps these GAPS.

Studies presented here suggest that GAPS can be reduced by rethinking some long-held customs and cultural influences about approaches to gaining an education. "People are strongly compelled to behave in ways that fit cultures." (Emerson and Stewart, *The Learning Developing Book*; 136) When it comes to meaningful learning, some long-held customs and cultures are suppressive.

The brain we all came into the world with is consciously and non-consciously evolving and encoding information from our experiences throughout our lives. IT IS THE HUMAN BRAIN THAT SPECIALIZES IN TURNING INFORMATION INTO ACTIONS. The view here is that all people start out at the head of the class and then, unfortunately, because of GAPS, some people are caused to fall behind when the culture of the approach to learning was not keeping "the brain in mind."

In his book, *Concept of Mind*, Gilbert Ryle noted that "knowing how" (i.e. application) is different from "knowing that" (i.e. details). He pointed out that there are forms of knowledge within the brain including practical knowledge, knowledge, knowledge by acquaintance, knowledge about objects, self-knowledge, background knowledge, and scientific knowledge. He also pointed out that there are emotional states of being careful, sensible, skillful, wise, prudent, clever, rational, stupid, and intelligent.

"When one decides to take learning seriously enough to do it as efficiently and enjoyably as possible, then one must take the brain's connection to learning into consideration." George Elliott, PeopleSmart.com

IT IS MENTAL

In a student-centered/brain-compatible environment for learning students will: question assumptions, seek reasons, be reflective, make connections, draw conclusions, learn from experience and observations. In these environments, students contrast, evaluate, and express personal perspectives. In student-centered/brain-compatible environment students also use self-reflection, foster creativity, and are comfortable because it is a safe learning environment.

Student-centered/brain-compatible approaches to learning change student's insights.

Your Brain at Work, by David Rock, CEO Results Coaching Systems, offers useful insights about the brain's connection to learning and performing including:

The less information we consciously try to hold in our brain the better.

Thinking about ourselves activates our default network and slows learning.

Useful insights arrive in a quiet mind, free of specific data.

Once an emotion kicks in, trying to suppress it makes it worse.

Putting a word on the emotion improves performance under pressure.

Noticing emotions and accepting them as they arise will reduce negative arousal.

Expectations will alter the quality of the information your brain receives.

Sorting out problems is not as useful as attention on what works.

A positive state of mind is reappraising not fixing.

Safe connections from instructors are vital.

The sense that our status is going down in the eyes others activates our threat circuitry.

Playing against yourself increases your status and self-worth.

The slightest feeling of improving can generate a helpful emotional reward.

Attention can go quickly and easily to a threat, like looking for what's wrong?

Once attention goes away from a threat, you create useful questions.

Be aware of your emotional state when you want to facilitate change.

Use solution-based questions.

"The Physical Reality of Being Human" (Eastman School of Music)

You will learn lessons.

You are enrolled in a full-time informal school called LIFE. Each day in this school you will have the opportunity to learn lessons.

There are no mistakes, only feedback.

is a process of trial and error. The failed experiments are as much a part of the process as the experiment that eventually works.

A lesson is repeated until learned.

A lesson comes to you in various forms until it is learned. Once you have learned, you then have what it takes to go to the next lesson.

Lessons do not end.

There is no part of life that does not contain its lesson. If you are alive, there are lessons to be learned.

There is nothing better than now.

What you make of your life is up to you.

We have all the tools and resources we need. What you do with them is up to you. The choice is yours.

Your answers lay inside you.

Your answers lay inside you. All you need to do is look, listen, and trust.

You will not be conscious of all you are learning. Learning is mostly non-conscious.

You can unconsciously remember whenever needed. Things come to mind without effort.

A MINDFUL EDUCATION

A student does not exist without a teacher, nor a teacher without a student. They are connected in a journey of cooperation. Modern science is continually discovering how miraculously interconnected everything is. Positive, playful learning, teaching and performing qualities arise naturally in a student-centered/brain-compatible learning environment by applying mindful practices to education. Mindsets before skills sets lead the way.

In *The Way of Mindful Education*, by Daniel Rechtschaffen, he noted that the aim of mindful education was to cultivate well-being in teachers and students. It turns out that the principles of a MINDFUL EDUCATION and student-centered/brain-compatible approaches to learning are connected. Much of what follows below comes from his must-read book.

Rechtschaffen expressed that a mindful education supports paying attention to the sensations in our body and mind (fear, nervous, relaxed, etc.) while staying aware of the surroundings. Once students learn to use their attention mindfully during difficulties they can bring themselves back to being connected to both the environment and their emotional selves, moving back into a safe place to learn and perform.

Mindful education approaches do not see outcomes as good or bad. With this method, our attention is rooted in the present moment without our minds getting involved in judgments. When mindfulness exists, we are thinking and labeling less and are open to what is there without judgmental labels. For example, a motion is just going high or low, left or right and is never labeled or judged as good or bad.

A mindful education starts with being aware. Are we relaxed or stressed: thinking a lot or not, breathing fast or normal, moving fast or slow, trying for outcomes? It is useful to be aware of how we are reacting (judgmental, or not) to what is going on in our environment, and the kind of chemical messages our bodies, minds, and hearts are sending throughout our central nervous system.

Mindful approaches to learning have been in play for thousands of years supporting cognitive development, attention skills, and emotional balance. When playing sports, making music, or creating art, conscious thinking often seems to move into the background. Mindfulness at its best is when we are totally and non-consciously absorbed in the present.

In student-centered/brain-compatible learning environments we are experiencing a mindful education and not memorizing someone else's view. We are setting up conditions in which we can observe an unlabeled experience in our own minds, bodies, and hearts. These moments of true awareness often arise spontaneously. We are mindful, open, aware, and seeing in a way that supports meaningful learning.

When the mind is on high alert, filled with self-critical thoughts, our working memory operates poorly; our creativity does not respond, and our collaborative skills are suppressed. On the other hand, when students feel safe, relaxed, and engaged in mindful unlabeled-attention, learning and change come naturally.

> ***The premise of a mindful approach to education and student-centered/brain-compatible approaches to learning is that all human beings come into the world predisposed with the skills that are the most beneficial universal qualities for surviving, developing, and thriving.***

The most efficient path to gaining an education is to approach learning that fosters the latent non-conscious qualities of creativity, integrity, and wisdom.

Mindfulness starts by accepting students and the outcome of their efforts exactly as they are, (workable or unworkable where nothing is broken or in need of fixing). With this type of attention, students can develop to their fullest potential.

Brain studies show that a mindful approach to development can positively transform the architecture and operation of our brain, improving sustained attention, working memories and concentration. Where negativity exists, the cost is little growth, learning, or development.

Joe Campos, of The University of California Berkeley Infant Studies Center, created an experiment in which some mothers were told to look at their child with total confidence and smiles, and other mothers were told to keep their faces cold and blank as their child walked toward them. The children were walking on a path in the direction of a visual cliff that had a glass cover over it. Those children whose attention was on their mother's confident smiles continued walking on the path when they came to the drop off that had a see-through covering. Those children whose attention was on the blank, cold look, on their mother's face stayed frozen at the edge of the visual cliff and did not keep walking.

Everywhere humans have lived, we have needed positive, mindful attention for hunting, making tools, or intellectual pursuits. Being aware of our thoughts and our feelings about what is going on in our lives rewires the brain and turns damaging reactions into constructive and positive ones. Everything is based on the story we tell ourselves. Developmental neuroscientists have studies that show mindfulness (accepting without labels) offers students key ingredients to experiencing healthy and fruitful lives.

As an instructor, I've always wanted students to have a path to their inner resources that turns their attention inward to their thoughts and feelings. What is going on in the mind and body? This is easier to do than trying to get the world around you to calm down. Shine a light of awareness on information, without the labels that experiences too often have.

Bringing the skills of mindful awareness to what seems difficult and causing harmful feelings is a tool that a brain-compatible learning process develops. Notice how the mind is always trying to busy itself with labels or something other than what is happening right now.

> ***After getting lost in harmful distractions, mindful awareness brings one back to the present moment. To accomplish this, we may have to view approaches to learning differently than in the past, leaving behind "teaching-fixing to get something right."***

There is no reason for beating yourself up for mentally wandering away. The fact that you have noticed your mind wandering means you are already paying attention, which is the good news. The point of mindful attention is not to get rid of our thoughts or create a state of total bliss. The intention is to notice what is true, with no labels or judgments. "Daydreaming is a normal human process." (Professor Terry Doyle)

When we get angry, sad, excited, anxious, or any other emotion, we usually begin to think a lot and get trapped in a mental loop, either reviewing what happened in the past or wondering what we will do in the future. Instead of being consumed by thoughts, we can mindfully cause a change of direction and accept what is there without judgments or labels.

When the information-delivery systems are not informed with an adequate understanding of how interconnected and interdependent the mind, brain, emotions, and physical actions are, they can disempower students.

We do not want to turn insights about mindfulness and the brain's connection to learning into another subject that students will stress about getting right. Students should feel they are not receiving information about a topic but are being invited to make their own discoveries and take something from what is being offered. The aim is personal authenticity, not behavior modification.

Students should be made aware that they will make mistakes while learning about mindfulness and not to question that reality. Discover how much you can learn from unworkable outcomes. Our willingness to learn more about a path to meaningful learning is more useful than just learning information.

> ***Insights (ideas) about information and learning outcomes are internal, invisible, and weightless, with no external physical substance. Insights have the power to make changes, or improvements, and enhance what exists. Ideas are able to prevent changes and suppress improvements.***

Tennis great Arthur Ashe said, "Start where you are; use what you have; do what you can."

After 40 years of teaching, the following insights came from Dr. Stephen Yazulla

"Today, with the emphasis on test scores, the assumption is that a failing school or student is the fault of the teacher. What is lacking in both "blame games" is the role of the family, support at home. A home environment that encourages education is likely to be every bit as important as the school environment in fostering student success. Students today are no less intelligent than those of the past. Teaching methods to-get-it right are also well entrenched and have not changed appreciably over many decades. So why are test scores across grades, reading, and math lower than in most other developed countries?

A rampant anti-intellectualism that pervades today's society that revels in ignorance are disputing scientific data in favor of myth and bias. There is much debate as to the cultural influences that foster an attitude of casual ambivalence to education. The suggestion to focus only on teachers' instructional methods ignores the 800-pound gorilla in the room–namely the home attitude towards education in general and science in particular. Cooperative support and social trust among students, teachers, and PARENTS are critical in any educational endeavor.

Often there is a big deal made of class size and teacher-to-student ratio. The smaller classes in grade school, high school, or lab courses are more conducive to the non-traditional methods alluded to here. During school budget and tax-related hearings, there is always the threat of larger classes, with the implication that education will suffer by diluting teacher time per student. The successes of tutoring and mentoring in minority programs or athletic scholarships are also due largely to the personal touch and individual attention.

However, in large universities, introductory courses in Biology, Psychology, Chemistry, and Physics often have several hundred to a thousand students. Here, a playful, creative, personal environment can break down. Lectures become a performance; exams and grading are more an exercise in crowd control. Nationally, in state universities, there has been an increase in student enrollment with a continual decrease in state support. Politically, resistance to tuition increases is offset by decreases in faculty/staff. It is important to realize which, and differentiate among the scholastic and other environments in which these non-traditional personal methods described here can be applied – that is, in small groups."

Dr. Stephen Yazulla

GOOD QUESTIONS

"The formulation of a problem is often more essential than its solution, which may be merely a matter of mathematical or experimental skill. To raise new questions, new possibilities, to regard old problems from a new angle, requires creative imagination and marks real advances." (Albert Einstein)

What follows is a summary of the fall 2010 issue of *Thought and Action*, a National Education Association Journal, wherein:

Professor Matthew H. Bowker and Evelyn Wortsman Deluty discussed the value of students asking questions.

Robert Zemsky and Lisa A. Sheldon discussed how to help students learn.

Joyce Lucas-Clark discussed that research could be demonstrated objectively in the concrete factual value of reality.

Thomas Axter discussed the future of education.

Kristen Dierking discussed keeping approaches to learning interesting.

Matthew H. Bowker pointed out that in the past we focused on how teachers should ask questions and how students answer them, overlooking the value of helping individuals develop their own questioning skills when learning. Bowker said, "if teachers are always ready to ask a provocative question their students would not have the need to take on that responsibility."

Bowker noted that several studies have shown that individuals demonstrate greater thought, complexity, initiative, and personal engagement when teachers do not ask questions, but instead state propositions or offer non-question alternatives.

Joyce Lucas-Clark pointed out that presenting facts or answers without developing questions produces the kind of environment in which everything is already settled. The primary objective of meaningful learning, according to Clark, is to improve students' ability to ask insightful questions, using answers as stepping stones from question to question.

Bowker wrote that in Marshall McLuhan's view when the teacher is the one who constructs the most interesting questions, problems, or critical challenges, students become dependent upon the teacher to create inquiry. On the other hand, a question-centered approach that is based on students' questions develops students who engage in course material as independent thinkers.

According to Bowker, requiring students to create their own questions helps them understand how answers are connected and beg for additional questions. In this environment, the questions themselves are the answers.

Bowker: The important value of asking questions is the ability to abstract from things; to unlock their meanings, courses, and consequences.

Bowker: Questioning involves speculating about possibilities both real and unreal, given and hypothetical, which is an immensely creative act. Questioning requires that the object not be seen just as it is.

Bowker: Questions probe to find something that is not already recognized, thereby discovering relationships and possibilities that are not yet given.

Bowker: In learning environments, questions must keep pace with answers. Both questions and answers must be appropriate to the levels of experience and cognitive development and functioning.

Bowker: Some believe that the purpose of education is to store up definitive answers in one's mind, but this does not promote the development of reasoning and deductive skills that create flexible knowledge and portable thinking.

Bowker: McLuhan pointed out that our capacity to generate answers is often less important than our ability to question answers we already have.

Bowker: Many instructors aspire to a state that a problem receives a perfect answer, which is not conducive to the development of a creative mind. It deprives individuals of the need to explore on their own, as they question more independently, revising answers.

Bowker: The relationship between the instructor and student must be caring, equitable, and responsive. It must be firm and safe. The tone must be playful and creative so students can think, converse, listen, and question without feeling either lost or crushed.

Deluty: Inquiry is the pivotal skill in the demonstrative process of meaningful learning. It is trying to judge for one's self rather than simply learning facts. Asking questions can transform one's mind from a passive recipient to an active participant in the process of deliberation and learning.

Bowker: Often there is a problem with a Socratic-based approach to learning when the teacher takes control of most or all the inquiry.

Bowker: The teacher who stays on top of students, one who is always the questioner, one who squeezes work out of the students, may produce students who have stored up the material, but not students who are critically learning. "One can only judge critically through practice."

Deluty: Learning to judge for one's self takes place when the act of transmission is not merely intended to impart facts, but rather to awaken students' questions.

Deluty: There can be an interactive process that emerges between the knower and the known. When this is applied to teaching and learning, the results shy away from a fixation on the content of the curriculum and, instead, are intended to draw students into the process. Palmer recognized that learning starts when individuals reflect critically, not when they are forced to remember facts.

Deluty: When approaches to learning evolve through questions from students, students are called upon to tap into their inner (non-conscious) resources and become participants in the process of learning rather than just bystanders.

Deluty: Questioning helps students leap across the gap that separates passive learning from animated participation in the transmission of knowledge. Questions are a sure sign that individuals are thinking critically.

Deluty: The mere assemblage of facts, no matter how great, is of no worth, without the habit of reflective inquiry to judge them.

Deluty: The ability to ask a reflective question is the root of all change and progress.

On the topic of inquiry, Wilhelm stated, "The research base is clear: inquiry-oriented classrooms cultivate motivation and engagement, deepen conceptual and strategic understanding, produce higher-level thinking, develop productive habits of the mind, engender positive attitudes towards future learning."

When Sheldon talked about motivation, she said her goal as a teacher was to facilitate student learning, promote goal attachment, and encourage academic momentum. Sheldon pointed to the value of intrinsic motivation. She said, "What separates successful students from the less successful is their ability to navigate obstacles and maintain motivation towards their goals." Getting derailed is a common problem that is actually positive.

Sheldon felt the essence of motivation is found in these two acronyms:

OARS and FRAMES

OARS = **O**pen-ended questions, **A**ffirmations, **R**eflective listening, **S**ummary statements

Open-ended questions allow students to tell their story. Open-ended questions cannot be answered with "yes" or "no" which is close-ended.

Affirmation – morale-boosting statements that help build confidence and feelings of improvement in students.

Reflective listening – listening and understanding what students say without judgment.

Summary – for students, it is a chance to provide additional information and be a reminder of action that needs to be taken. To keep students curious is to keep them motivated.

FRAMES = **F**eedback, **R**esponsibility, **A**dvice, **M**enu, **E**mpathy, **S**elf-efficiency

Feedback – A clear, non-judgmental way that articulates the current behavior and the goal of new behavior.

Responsibility – Behavior is owned by the students and therefore it is up to them to change it if they need to.

Advice – Provide new perspectives or options to consider; it must be the student's choice.

Menus – Offer more than one selection to a problem and allow the student to choose.

Empathy - Acknowledge that change can take time and work.

Self-efficiency – Expressing a belief that the student has the potential to be successful increases the feeling of improvement in students.

Sheldon suggested that motivational conversation places responsibility for change and success in the hands of the student. Avoid confronting change head-on as it only feeds resistance, which then becomes the root of the problem, and not the behavior itself.

Barbara Fischer Ph.D., founder of United Clinical Services, said to avoid seeing students as consumers to give something to and help them gain something. Create environments and have ideas mingle and give rise to new knowledge (know-how skills). Develop environments where more is being heard than is being said.

When Professor Joyce Lucas-Clark talked about science, she said it appeals to human emotions and is characterized by enhancing uncertainty. Science is supported by gathering information and experimentation (sounds like how we learn anything).

Professor Randall E. Jedule said, "When I learned about active learning, it was as if the goddess of education waved a magic wand and turned me in a new direction. I devoted my doctoral research to how teaching might be altered through cooperation."

Professor Kristen Dierking, the recipient of NEA's Excellence in the Academy Award for the Art of Teaching, stated that approaches to education must move beyond simply feeding facts to students without asking them to think creatively or critically about the material. "I did not want to just dish out information; I wanted to provide content in a way that would help students retain material. I began to use real-life examples when appropriate to illustrate concepts as clearly as possible. I often found that once you agree on a viewpoint, you're likely to forget it. I try to have students debate their point of view."

"I want students to incorporate what they learn into the body of knowledge that they can carry with them, thinking in a diverse manner. I spend considerable time encouraging students to examine why they think what they think. Asking students to say or write something organized has produced many new insights and comments. It helps to know the concept behind a student's organized work."

"I encourage creative risk-taking. I emphasize how important creative work is and how necessary their kind of thinking would be in the student's future lives. Building a student's confidence in his or her ability to innovate is invaluable. I try to help students to see the world in a way they had never previously considered."

KNOWLEDGE TO GO

What follows is based on a lecture by David Perkins, Ph.D., Harvard University Connecting Mind-brain to Education Institute on the topic of Knowledge to go.

Ideally, knowledge is not something we give to each other. It should be seen as knowledge to go: knowledge to think about, knowledge to think with, and knowledge with which to see and act.

Go from knowledge about information to acting with information.

Learning is making something more out of something basic.

Personal knowledge is what will be retrieved.

Find what makes information matter to students, eliminating prior perceptions that can interfere.

Meaningful Learning: learned in one area and can be used in different areas.

Passive Conceptual Vocabulary: (non-active, not personalized) is useless, but can be helped by questions.

Often answers and outcomes have been over-thought and under-learned!

If you look at an iceberg, some ice is above the water, but most of the mountain of ice is below the waterline. The ice below the water keeps the iceberg in balance.

Imagine the ice below the water as our non-conscious mind that holds a large volume of information made up of our experiences that were non-consciously encoded for future use and non-consciously influencing new learning. New learning is supported by prior knowledge.

We must move on from threshold ideas and look outward into the world, but not for details. Be creative. Start with an idea; see it as a tool, and then use it.

The brain is a self-organizing system.

Within the brain, learning changes how genes express themselves, influenced by experience and environment.

Every time our eyes move new information arrives, changing what we perceive. Our perceptions have implications for learning.

Perceptions are about survival; they are a generalization; a variable that begins with the eye and is then constructed.

The brain pays attention to change. The brain non-consciously sees edges and contrasts automatically. The brain is not interested in steady-state conditions.

Black and white pictures are encoded faster than color pictures which provide less personal information.

What we currently know shapes our perceptions. This is also known as "bias."

IT IS EMOTIONS

that judge

that criticize

that want something

that have perceptions

that have beliefs

that change our mind

that gets us upset

that tell us stories

that gets us afraid

that have doubts, cause fear & thereby

SUPPRESS LEARNING

RECAP

The nature of learning is similar to museums where one is guided by their own curiosity to have interesting interactions. Interactions are one of the nature-of-learning's most useful tools, supporting active access to new insights and development.

Meaningful learning comes out of us, not to us, based on our retained experiences.

Meaningful learning is more of a product of wide observation than the result of directly focused thinking.

Some traditional approaches to learning that do most of the thinking for the student do not help students learn.

When discussing Learning, include the elements of memory, emotions, organization, context, environment, experiences, the process being used, and the brain. Learning is a team effort.

In a student-centered/brain-compatible environment for learning, students will question assumptions, seek reasons, be reflective, make connections, draw conclusions, learn from experience and observations.

Contrast, evaluate, and express personal perspectives in student-centered/brain-compatible environments. Encourage self-reflection and foster creativity because both are frustration free and consistent with an emotionally safe learning environment.

A question-centered approach to learning is based on students' questions (which in and of themselves is the answer). This approach develops students who engage in course material as independent thinkers.

We must move on from threshold ideas and look outward into the world, but not for details. Be creative. Start with an idea; see it as a tool, and then use it.

YOUR NOTES ____

____

____

CHAPTER SIX

Studies, Strategies, and Research

Harvard's Connecting Mind-Brain to Education Institute, UCLA's Learning and Forgetting Lab, Vanderbilt's Peabody School of Education, The University of Washington, The Human Brain Project, National Institutes of Health, Allen Institute, Salk Institute, and many other universities and research centers are investing billions of dollars to inform educators about the human brain's connection to learning that should be taken advantage of.

AN INVITATION

It was an honor to be invited by Dr. Debbie Crews to speak during the World Scientific Golf Conference at Arizona State University, March 2001 and again in 2016 at St. Andrews University Scotland. The aim of these prestigious conferences was to influence and improve learning in the future. What scientists and educators from all over the world presented about teaching and learning that week was relevant for learning anything. To be in the company of these men and women was invaluable. Some of the research presented at these gathering included the following:

Random Training supports long-term retention of skills more efficiently than drills and rigid blueprints. It enhances learning to change environments frequently.

Self-learning, or education by selection, or by choice, is more effective for long-term retention than following directions.

A learning model that is general in nature is more effective for retention of skills than a specific expert model. Models that are just in-the-ball-park lead students in the direction of highly skilled results more efficiently than expert models that are better used as inspirations to motivate.

> **Efficient learning environments pay attention to the learner, not poor habits: help learners discover what's needed, provide learners with the opportunity to personalize information, develop the tools of adapting (reverse engineering).**

A number of factors that can fragment learning were discussed including: stress, rigid systems, drills, rote learning, expert models, lack of core knowledge, little or no creative play, little sensory stimulation, little opportunity for developing imagination, few chances to solve problems and training in a get it (or did not get it) environment.

Columbia University

Drs. Bailey, Hughes, and Moore (2003) pointed out in their book, *Working Knowledge, Work-Based Learning and Education Reform*, that today cognitive scientists are drawing on a wide range of sources including some 19th century educational thinkers to add to their own discoveries about how people learn. For example, in 1875 Francis Parker replaced a traditional teaching curriculum with educational projects and learning experiences that were meaningful to students in schools in Quincy, Massachusetts.

One of America's first great thinkers about education and founder of progressive education in the United States, John Dewey, built on Francis Parker's insights about learning. Dewey saw traditional education as being isolated from reality and passive in its methods. The approaches to learning for schools that Dewey created at the University of Chicago had three principles.

1) Instruction must focus on the development of the student's mind, not on blocks of subject matter.

2) Instruction must be integrated into project-oriented approaches.

3) The progression through years of school education must go from practical experiences to formal information to integrated studies.

Dr. Bailey also points out that John Dewey's principles required students to make observations and predictions, thus developing the student's own scientific skills. Regarding designing efficient learning environments Dewey left us some key ideas:

1) The student is the center of learning.

2) Learning is an active engagement, with an environment structured for education (not test scores).

3) Integration of mind and action, head, and hand, academic, and vocation promotes learning.

4) Learning takes place in context (real world situations).

5) Learning is guided; providing structure for connections between every experience.

6) Learning in context gives students insights about the role of information in problem solving. Drills often are not in context; copying expert models is following, which is a lower brain activity. Copying is not problem solving. Concepts and skills are acquired as tools, with a range of problem-solving purposes for enhancing flexible knowledge and portable skills.

Dr. Bailey and others spoke about the value of apprenticeships (paraphrased below). "The success or failure of a task that has been performed is usually obvious and needs no commentary. The person who judged the apprentice's performance was the apprentice himself or herself rather than the expert." (Dr. Bailey)

1) Traditional apprenticeship programs in the 19th century were a source of ideas, promoting efficient learning environments for how individuals learned. Spending time in apprentice programs was not so much practicing for the real thing, as it was doing many useful and necessary tasks. Apprentices were driven by the work to be accomplished (not how-to information).

2) Apprenticeships should not be recognized as a conscious learning effort at all. Much of what an apprentice did is hard to differentiate from play. Traditional apprenticeship programs used apprentices for trial runs and practice, whereas, schools tend to be a practice of specification.

3) During apprenticeships, the evolution of a learner's competence emerged naturally and continuously in the context of work and was evaluated by what was being accomplished, rather than from a test.

4) After the apprentice was aware of core knowledge, he knew what remained to be learned. The ability to move on to the next skill was largely under the influence of the apprentice rather than the master. In other words, the apprentice tended to own the problems of learning and was in charge of moving on to the next skill.

5) During an apprenticeship, whatever instruction the apprentice received originated not from a teacher doing teaching, but from the learner doing, observing, and adjusting.

6) Traditional apprenticeships showed what efficient context-led learning looked like. Cognitive science has developed analogies appropriate for learning through the apprenticeship approach based on playful self-discovery. When students are being told how to do something, it is a violent change from nature's plan for self-development with external learning influences that should be avoided at all costs.

Efficient approaches to learning could easily be called interactive brain-on, hands-on entertainment, as students invent their own style or skills and develop their own information base, (or what is referred to as their own working knowledge).

Mankind's brain is designed to distil, translate, and interpret; not passively follow directions. Joseph Clinton Pearce, the author of *The Magical Child*, described this design as a "two-way flow of assimilation and accommodation."

INNOVATION AND LEARNING

Creating Innovators, a ground-breaking book by Tony Wagner, is a must read for everyone who has an interest in the topic of learning.

Tony Wagner is an Education Fellow at Harvard's Technology and Entrepreneurship Center. He researched what supports and what suppresses innovation and learning. Wagner conducted over 150 interviews with students and their parents, teachers, and professors, engineers and scientists, artists and musicians, entrepreneurs and corporate C.E.O.s, and military leaders. The following are some of the answers Wagner received when he asked the interviewees to define “innovation;” innovare (Latin); to make new into; renew, change.

“I do not define it technically because it is an art. Innovation is the process by which new things take place. I look at innovation as an approach.” (Dean of London Business School, Sir Andrew Likerman)

“Innovation, the process of having original ideas and insights that have value; then implementing them. It is easier to name what stifles innovation and how to kill creativity – like rigid structure and high stress.” (President of Olin College of Engineering, Richard Miller)

“Innovation does not have to be about creating the next iPad. It can be the way you treat a customer.” (Mr. Joe Caruso)

“Innovation is creative problem solving.” (Proctor and Gamble’s Director of External Relations, Ellen Bawan)

> **Innovation and learning come in many forms. There are interchangeable insights that run through the common roots of both. When innovation and learning are meaningful, Wagner often found non-traditional approaches are used for traditional problems.**

Kirk Phelps, a successful innovator and master learner, noted his learning environment was mostly non-traditional, supported by parents who were comfortable with a trial-and-error approach that did not include criticism and judgment. As a child, his parents did not care if he was on a winning team or not. They just wanted him to experience different kinds of people and develop an interest in sports. Wagner states, “You cannot innovate from nothing. You must have some knowledge, though how much knowledge you need, when you need it, and how best to acquire it are important questions.”

Kirk Phelps told Wagner, “What you study is not that important. Knowing how to find things is way more important (integration on a personal level).” What tools did I want to add to my toolbox? My parents did not care all that much about what I was interested in; they were far more interested in the process of experiencing learning. Explore, experiment and discuss through trial and error–to take risks and to fail without criticism supports learning.

M.I.T.’s Joost Bonson told Wagner, “Being innovative is central (natural) to being human – we are all playful and curious animals until it is pounded out of us.” How true this is when approaches to learning are not brain-compatible/student-centered! Reflecting on his research about meaningful learning, Wagner said, “I now understand that the qualities of innovation that are all elements that outweigh subject content: Perseverance -Willingness to experiment -Take risks -Tolerate failure - Design thinking -Critical thinking.”

In another of Tony Wagner’s books, *The Global Achievement Gap*, he suggests seven survival skills for continuous, lifelong learning as:

1) Problem solving

2) Calibration

3) Adaptability

4) Finding and analyzing information

5) Intuition

6) Effective oral and written communications

7) Curiosity and imagination

The elements on Wagner's two lists are all found in brain-compatible/student-centered learning environments, as are the skills that Jeff H. Dyer, Hal B. Gregerson, and Clayton M. Christensen wrote about in their article, The Innovator's DNA, (Harvard Business Review, Dec 2009). They divided these skills into two categories-Doing and Thinking.

DOING includes:

Question: Questioning, allows individuals to consider possibilities. Judy Gilbert, Director of Talent at Google, stated, "Of course we look for smarts, but intellectual curiosity is more important. We look for someone who questions how they can make something better. David Kelly, the founder of IDEO, a global design firm, has the slogan "fail early and fail often."

Observe: Observing detects new ways of doing things. Wagner points out the importance of play with his statement, "Observe, explore, imagine, and learn through play more than we would ever have thought possible." Alison Gopnik (University of California, Berkley) discusses the progression of play from passion to purpose; through opportunities to explore by trial and error–play is brain-compatible/student-centered learning.

Experiment: Experimenting tries new experiences to explore possibilities.

It is a poor concept to feel that all a model had to do was act in a certain way; attempts to imitate do not help learners; merely repeating someone else's conclusion is not experimenting. The joy of learning is based on experiencing all the stages (the ups and downs) of preparation. Struggle is nature's indirect approach and preparation for learning. Efficient preparation develops flexible knowledge and portable skills (innovation).

Preparation and struggle are two of the most critical stages of long-term progress and are often overlooked and undervalued.

Networking: Networking gains radically different views from diverse backgrounds. You cannot engineer innovation, but you can increase the odds of it occurring. Innovation, broadly defined, is the critical ingredient of learning. Learning- development approaches produce the raw material of innovation and the kind of learning environments that encourage diversity, experimentation, risk taking (not trying to get-it-right), and combining skills from many fields when learning.

Thinking: includes association which links patterns of doing, cultivating innovation and learning.

The culture of teaching to get-it-right rarely supports innovation. That culture lacks wide-ranging curiosity and support for continuous learning that is vital to growth. The opportunity to draw from a wide range of diverse experiences can generate many breakthrough ideas and insights.

Innovators are curious; they take risks, and they can associate concepts across databases. That is, they can step back and view a situation from a broad view, rather than be immersed in the details–the "forest and the trees" idea. Steven Jobs, perhaps one of the most innovative men the world has ever seen said, "Be curious, experiment, take risks, stay hungry, and stay foolish."

Walter Isaacson, the author of Steve Jobs' biography, said there was a difference between intelligence and genius and Jobs was more genius. He had imaginative leaps that were instinctive, unexpected and at times magical. They were sparked by intuition, not analytical rigor.

Isaacson said Jobs came to value experiential wisdom over empirical analysis. He did not study data or crunch numbers, but could sense what lay ahead. Jobs appreciated the power of intuition, in contrast to what Isaacson's, "Western rational thought."

Steve Jobs' intuition was based not on conventional learning, but on experiential wisdom. He had a lot of imagination and knew how to apply it. As Einstein said, "Imagination is more important than knowledge." Jobs was super ingenious and enjoyed the concept of applying creativity that spawned innovations that changed the world.

HARDWIRED

Nigel Nicholson's article "How Hardwired is Human Behavior?" contained insights about *Homo sapiens*. Many of our early ancestors' survival traits dating back to the African savannah more than 200,000 years ago, are still hardwired into our 21st century brains. (*Harvard Business Review*, July/August 1998 article)

That article has many insights and suggestions that are based on brain-compatible/student-centered information-delivery systems. Some aspects of human behavior, including the skill of non-conscious learning, are biologically inborn and universal. Nicholson used six sources of scientific research to uncover the influence of these hardwired traits. (Anthropology; Behavior; Genetics; Neuropsychology; Paleontology; Social Psychology)

We are hardwired to use emotions as the first filter of all the information we receive (a hugely valuable insight).

Learning Approach: Avoid criticism, judgments, and negative evaluations of unworkable outcomes.

We are hardwired to avoid taking risks when feeling relatively secure and fight when feeling threatened.

Learning Approach: People will act and think creatively when given support in a safe space.

We are hardwired to feel more self-confident than is justified.

Learning Approach: Help people to realistically evaluate the situation and challenges they face.

We are hardwired to quickly classify situations, people, and experiences into good-bad categories.

Learning Approach: Promote the value of objective evaluation. Something is what it is; not good or bad, easy, or hard.

We are hardwired to participate for status and ego.

Learning Approach: Encourage the engagement of the task for the sake of the task itself.

We are hardwired so that positive remarks do not erase the damage of negative remarks.

Learning Approach: Always be proactive; stay away from negative remarks.

We are hardwired to seek survival.

Learning Approach: Promoting the recognition of the result of a journey.

FEEDBACK

When long-term learning is the aim, some considerations that enhance learning include:

Using metaphors

Using real-world environments with more than one solution

Using more than one example to support findings

Using a variety of points of view on a topic - both workable and unworkable

Using comparisons and contrasts in different context

Phil Race, a Professor of learning and teaching at Leeds Metropolitan University, discusses "feedback" in his book Making Learning Happen. He mentions attitudes that support efficient learning. Race suggests that individuals must:

Recognize the need to learn.

Learn from trial and error.

Be able to make personal sense of information.

Learn from workable and unworkable information.

Learn from self-exploration and self-assessment.

The following statements are Race's suggestions related to Feedback:

Help individuals learn how to best use different kinds of environments as sources of information.

Help people see the big picture, not just details.

Provide overall information guidelines.

Help people understand the real meaning of results.

Provide opportunities for clarification.

Help individuals recognize life's differences and similarities.

Help individuals recognize strengths instead of weaknesses.

Race based his suggestions on studies into the brain's connection to learning and how emotions influence that connection. At times, the feedback we receive is a non-conscious message.

David Brooks' book, *The Social Animal*, is a thesis on how human beings are influenced by the hidden workings of our non-conscious minds.

Unknown to our conscious self, the power of deep, prior knowledge from our unconscious shapes the way we think and learn. Avoid the trap of thinking that decision making requires access to much information. Allow your non-conscious to kick-in and remove your conscious thinking self from the process. The mind often works best when it ignores some information.

"Intuitive, creative, design-making relies on distilled experience. More information can overwhelm the brain, making it harder to tap into core information." (Eric Kessler, Pace University)

THE EVERYDAY GENIUS

In Michael Alexander's preface to Peter Kline's book, *The Everyday Genius*, several insights capture useful suggestions for enhancing acts of learning. Alexander starts by pointing out "despite the billions of dollars and millions of hours invested in education, for many students learning is often deferred." He went on to say, "This is a tragedy of enormous proportions, with terrible implications, but this does not have to continue."

For the past three centuries, we were told that what teachers were doing was correct, and we blamed students who failed to learn. Recently, rather than blaming students for not learning, researchers have been exploring the who, what, why, when and the where of learning, pointing out that a teacher's every move supports or detracts from the learning process. The total environment must be orchestrated to support learning.

What follows is based mostly on *The Everyday Genius* as a review of "integrative learning," based on research from Georgi Lozanov and others dating back to the 1950s.

The principal hypothesis of *The Everyday Genius* is that by changing the learning environment and the way information is presented, we can get substantially better results than are possible with traditional approaches to education. Traditional methods of teaching almost always fail to match the way we naturally learn.

The real cause of the problem is the culture of the system. Many students are hampered by outdated concepts of learning that often bring out the worst in both teachers and students. This does not mean throwing out everything and starting over; it only suggests making better use of the new resources that are available now.

> **Maintaining the status quo will not benefit anyone, and fortunately, changes are now available for improving approaches to learning. The reality is that it is important to widen the circulation of information about optimum conditions for learning and how they can best be implemented.**

The aim of integrative learning is to learn like children do. Although some new suggestions for enhancing acts of learning may seem childlike, they have been used with adults, and the learning was dramatic. The same learning activities used by pre-schools were used by corporate executives.

The nature of learning and the brain's connection to this process of growth is so basic to the human experience that it has meaning at every level of the developmental process. Most learning difficulties stem from ignoring this fundamental principle. Integrative learning is a seamless approach that recognizes that the development of the whole person must not be sacrificed for the development of a skill or intellectual progress. Some approaches to learning are discouraging, which disempowers the student/learner.

Learning without unpleasantness does not suggest learning without experiencing challenges and unworkable outcomes. Integrative learning recognizes the brain's connection to learning, and therefore how important unworkable outcomes are for experiencing meaningful learning. With a deeper understanding of the nature of learning, unworkable outcomes are no longer seen as a failure; they become valuable feedback for future reference.

One of the keys to meaningful learning is to adapt the approach to learning that is used to the way we naturally learn. To educate means "to lead out from." Recall that Benito Mussolini, the Italian dictator, was referred to as "Il Duce," the leader. The task is not to impose learning, but to lead learning out from resourceful unconscious minds filled with the memories of workable and unworkable outcomes. Many approaches to learning stop too early and do not utilize brain-compatible/student-centered learning environments to encourage students to use their unconscious minds that hold what is waiting to be expressed beyond the mere awareness of information.

During childhood, lots of learning occurs that is beyond awareness. Learning should always be the most natural and entertaining activity that adults and children experience. No emotional unpleasantness allowed! Emotions are at the center of everything we do or don't do, including learning.

Criticism, no matter how well intended, is never as supportive or productive as positive, patient encouragement. The put-downs and negative evaluations, too often found in traditional teaching, work against optimal learning experiences. Instead of griping about the current situation or outcome, consider how things might be different. Self-initiated positive behavior can build the necessary conceptual base for what-to-do, not what to fix when learning.

High expectations are valuable, but they should not be confused with specific outcome goals and predetermined milestones. Find challenges that lead to growth and eventually to the individual's full potential and capability. Outcome-goals tend to be specific; process-goals involve staying in the present. Avoid outcome-goals, while using process-goals! Instead of saying "I will break 90 or 100 in my golf score today," pay attention to the opportunities that each golf shot provides. Eventually, the golf score will improve.

When providing new information, or reviewing past experiences, meaningful approaches to learning encourage and reinforce the student's belief in their childhood capacity to learn and succeed. The emotion of fear must be banished from approaches to learning and replaced with the constant emotional support of the student's self-worth.

As far back as the 1950s, Georgi Lozanov, (who was known as the father of accelerated learning) and other researchers came to recognize and believe that everyone had the potential to learn at a rate two to five times faster than the present education system considered normal. They also believed that even that rate of learning could be improved.

> **Lozanov proposed learning environments be childlike playrooms, where instructors take on a nurturing role, reinforcing the learner's own attempts to learn with positive responses. Downplaying and looking past unworkable outcomes and seeing them as useful attempts, is a component of integrative learning.**

Lozanov called this approach "suggestopedia." This implies that general suggestions, not how-to directions are the cornerstone of information-delivery systems. This approach points to self-actualization and success orientation (not fixing). Humans are not wired to fail tests, miss golf shots, or have car accidents; we are all wired for survival. This means that we try, adjust, try again and again. We adapt and survive.

STRATEGIES

Some strategies and approaches support personal innovation, self-reliance, and long-term learning. Some strategies for finding points of entry for modernizing learning include:

> Attention on what to do, not how to do it or fixing outcomes. Learning is based more heavily on outcomes that are not criticized or judged, than on workable outcomes.
>
> Create safe non-judgmental learning environments.
>
> See all outcomes as feedback for future reference, not as a failure.

Failure = **F**inding **A**ccess **I**nto **L**earning, **U**ncovering **R**elevant **E**ducation

Students do not need evaluators–they need a guided search for solutions. Use smart approaches that realize no student is broken or in need of fixing. (On a journey of learning and developing, the present outcomes have nothing missing, they are just developing.)

Use a play-to-learn approach. (When incorporating new learning, use a playful, random approach.) Help people invent their own skills.

Giving students choices can be more useful than giving those directions or corrections. (Do you like A, B, or C?)

Let learning be an autobiographical process that promotes self-discovery and self-assessment.

Learning will be more efficient when how-to directions are avoided.

Promote self-regulation; make students feel "capable" by developing their self-reliance skills.

The use of broad, general concepts and metaphors can be more useful than using detailed information. The brain can use a general, just in-the-ball-park concept. Help individuals understand general concepts. The brain just wants the gist of things.

The following are notes I made while reading *Teaching for Understanding*, edited by Martha Stone Wisk:

Richard Meyer, "Conceptual models, promote understanding and solving problems more flexibly than when students do not develop a conceptual model." You can have a mental model of anything without understanding it.

Mental models involve a certain point of view, and while they are informed, they can also be misleading; Conceptual models avoid this contradiction.

Conceptual models explain, justify, relate, extrapolate, and apply information in ways that go beyond one's knowledge and routine skills.

Conceptual models can be simple, but rich with implications that support thinking and acting flexibly with what one knows, much like a jazz musician.

How ideas, processes, relationships, patterns, and questions are investigated forms the basis of conceptual models.

The subject matter is the sub-goal, with conceptual models supporting the process of learning and performing the main goal.

When learning is approached through a range of entry points, from various topic perspectives, while using a variety of learning materials and resources, conceptual models can be developed.

The framework of teaching-for-understanding is almost as old as human history. "Parables and metaphors that make connections with student's worlds have been used by teachers going back before Plato's time. For example, well-known fables attributed to Aesop date to before 550 B.C. Parables are also fundamental to ancient Buddhist texts and the Gospels of the New Testament. These stories construct images and allegories that facilitate understanding of simple, as well as obscure, concepts." (Professor Stephen Yazulla)

Teaching for Understanding **incorporates the view that learned information needs to be used in different circumstances in and out of the organized learning environment, thereby serving as a base for ongoing and extended learning that is always alive with possibilities.**

Educator, John Frederick Herbart suggested that each stage of learning had to be integrated with the past while looking for entry points that connected with current knowledge. This view was similar to Alfred North Whitehead's ideas of a continuous learning cycle.

When something has been learned, it can usually be demonstrated by some performance. What may be less evident is that what made meaningful learning possible was a brain-compatible/student-centered approach.

Michael S. Gazzaniga, one of the world's leading neuroscientists and author of *Human* wrote, "We humans are special. All of us can solve problems effortlessly. Humans have the ability to reflect on their own thoughts which are called metacognition, a component of human existence while interacting with the environment."

When utilizing *Teaching for Understanding*, puzzle solvers should be made to understand:

Their present ideas and feelings influence learning.

Learning is a multi-step non-linear process.

Learning plays a significant role in all our lives.

Learning compares and makes decisions about what to keep or reject.

Learning can be approached in such a way that it can empower individuals.

Learning for understanding includes comprehension, design, enactment, and integration.

For example, find key elements that students can comprehend and relate to, then design the curriculum on those key elements. Use an approach to learning that causes students to enact the key elements of the topic. Integrate those key elements with prior knowledge and experiences.

Learning for understanding relates directly to insights about process goals.

Avoid criticism while developing and applying opportunities for understanding.

Promote and engage multiple expressions of learning in different contexts.

Create reflective engagement in challenging, approachable tasks.

Provide opportunities for students to demonstrate their understanding.

Use implicit insights that move beyond explicit descriptions.

Meaningful learning produces flexible knowledge and transferable skills, and not just one view (a reflection of learning for understanding). What follows was adapted from *Teaching Today*, by Geoff Petty:

Research by Professors John Hattie, Robert Margorno, and others has demonstrated that cognitive and constructivist approaches appear to provide the most efficient paths to meaningful learning.

Principles of cognitive approaches to learning include:

Use higher-order skills, including planning.

Use high order tasks requiring analysis, evaluation, and synthesis.

Use laddered tasks that go from simple and concrete; from obvious to more abstract.

Principles of constructivist approaches to learning include:

Use approaches that require students to create and apply their own ideas to decide what to do.

Use approaches that support what the student is doing. Acts of teaching are just a means to the end; it is learning that counts.

Use a trial-and-error approach to learning that requires students to evaluate their own workable and unworkable outcomes as feedback for future reference.

Create student's participation that makes acts of learning interesting.

Keep in mind that learning is grounded in acts of doing in emotionally safe environments.

Common cognitivist and constructivist learning strategies include:

Asking questions and guiding discovery.

Have students identify key parts and how they relate to the whole.

Have students discuss the material to be learned and construct insights that are personal in nature.

"The ultimate goal of the education system is to shift the burden of pursuing their own education to the student." (J.W. Gardner)

***Teaching for Understanding* has a framework that puts attention on what students do, rather than on what teachers do. Understanding grows by performing one's current knowledge.**

Let's define "understanding" as; "going beyond the information given; to extend, synthesize, apply, or otherwise use what one knows in creative ways." This extension includes, but is not limited to exploring, interpreting, analyzing, relating, comparing, and making analogies.

A brain-based approach will utilize a full spectrum of intelligence and perform in myriad creative ways to aid understanding. Instead of rehearsing or recreating what others have done, true understanding engages learners in creating their own insights, which they can then incorporate into their own autobiographies.

The following are notes and insights from *How People Learn* (Bransford, Brown & Cocking, ed., 2000)

Begin with the student's questions rather than with a fixed curriculum. (p. 157)

Topic knowledge without knowledge of how individuals learn and how to guide the processes of learning would not yield meaningful learning. (p. 157)

A change in models of efficient learning is needed from what many teachers, parents, and students currently use. (p. 141)

Expose students to the major frustrations of the topic to be learned. (p. 139)

Engage students in cognitive conflicts and then have discussions about conflicting viewpoints. (Piaget, 1973)

A major goal of education is to prepare students for flexible adaptations to new problems and settings. Time spent learning for understanding has different consequences for transfer than time spent learning facts. (p. 77)

Rather than focus on specific solutions, highlight the general features of an action or critical decision. (White and Fredrickson, 1998)

Have students use self-assessment. (p. 140)

CAINE'S RESEARCH

When conducting research, Renate and Geoffrey Caine traveled all over the United States asking questions of many people. They point out that mental models shape our day-to-day decisions and interactions, and these models are often deeply ingrained with assumptions and generalizations that are inaccurate.

Mental models are the way we perceive the world and make sense of it. These fundamental beliefs are powerful because they help organize experiences, information, and strategies. The answers that the Caines obtained during their research reflected some of the finest thinking of our time.

They heard, "*Learning* is a change in thinking and behavior due to new understanding. *Teaching* is giving learners the opportunity to do their own best learning. *School* is anywhere this can happen, and it does not have to be a classroom."

The kind of brain-based research that the Caines used to gain an understanding was quantitative, participative, and dynamic, as are all the studies and research compiled here. The Caines said, "We observed, collected data, asked questions, searched for answers, read profusely, modified and expanded assumptions connected to our knowledge, and then we would start all over again. This is the same process to which we should add that we sought to induce in the people with whom we worked."

The Caines' research led them to a summary of three instructional approaches, with approach # 3 being the most efficient in their view.

Approach 1. This approach relies on top-down thinking and the control of information and facts to be disseminated by teachers. A stand and deliver method.

Approach 2. This is still primarily a command and control mode of instruction, but it is organized around concepts rather than just memorizing, with an eye to creating meaning.

Approach 3. This differs radically from approaches #1 and #2 because it is more learner-centered (from bottom to top). This kind of instruction is more fluid and open. It includes elements of self-organization as students had purposeful projects, critical ideas, and meaningful questions, within a context of educational experiences that approach the complexity of real life.

PLAY

Dr. Robert W. White, Professor Emeritus, Department of Psychology, Harvard University talks about "Play" in his insights that follow:

"It seems that each paper is written with a flavor that represents the biases and commitments of its author. Certainly, this paper is no exception. I would like to share a few of these with you.

I am a teacher. Although most teachers teach with books, pencils, and paper, I teach and learn through play, a tool that has continued to win my utmost respect and fascination.

I have discovered that play has certain very exciting potentials, but that these can come to fruition only when those who direct or affect the play experience view play and players in certain ways. I have tried to evolve some parameters that might begin to clarify this particular view of play and players."

The players are the most important part of any play experience.

Children have social, emotional, cognitive, and physical needs.

When skillfully planned, movement activities have the potential to meet important developmental needs.

Everyone has the right to healthy positive play experiences. It is our responsibility to find ways to make this type of play available to all.

An individual's view of self, others, and the world can be affected by play experiences.

Play behavior may be an attempt to deal with unmet needs.

What children learn, and what they become depends largely on how they feel about themselves.

We can become more sensitive to needs and more skillful at selecting, modifying, and/or creating movement activities to meet these requirements.

Present play activities need to be studied carefully in relation to both their positive and negative effects upon participants.

Movement activities/games can be developmental tools for both growth and motivation.

The process of play: **When play is used as a developmental tool, the process of playing** becomes the means by which players grow. The process may need to be guided. Is it a growth-producing experience for every participant? Within the process, there must be the opportunity for each player to find that s/he can overcome a challenge through effort. This may involve fumbling, generating of some ineffective alternatives, self-evaluation, and making adjustments. When play is used as a developmental tool, this process of struggle is the means through which an individual grows. Do adults need to support this process and avoid the tendency of giving too much emphasis (and praise) to who was the best, who won, or the perfection of athletic performance?

Thoughts to ponder: Should self-evaluation replace our present emphasis on praise? Don't we want a child to become an adult who can function effectively in solving problems and making decisions? Could we recreate play to allow for self-evaluation and self-correction?

It is hoped that you may find the excitement we have found in using the games as developmental tools.

Why use movement activities? "Play" is exciting and fun filled. Its elusive qualities draw and hold its participants' energies and concentration. There is virtually an inexhaustible supply of movement activities, which are flexible and can be modified or progressively changed to meet specific group needs.

Movement activities are action based and observable. Not only, the planner, but also the participant is receiving immediate and constant evaluative feedback. Just as one learns muscular control through frequent and repetitive experiences, so may play be a tool to evaluate social interactions and to experience and deal with various emotional responses and personal feelings.

It has been my experience that well-planned play increases a child's willingness to become involved, and in turn, more ready for the experiences that follow. Activities that allow a child to solve a problem, make a viable decision, and to feel personal success seem to increase a child's active efforts to cope and his or her willingness to take chances. This effect certainly goes beyond the historically acknowledged value of play as simply a means of letting off steam or reducing stress.

What is your role? Those who direct or affect children's play have a vital role in each child's growth and development. The awareness of the children's needs, careful selection and perhaps modification of activities, and continuous observation and evaluation can increase the possibility of play experiences being a positive contributor to growth and development.

PLAY, continued: Brain-compatible/student-centered approaches to learning create emotionally safe learning environments that are playful in their approach toward making students feel capable, which is perhaps the main responsibility of any provider of information. When people feel capable, they take the kind of risks and struggles that learning is founded on. All play is practice for life. Play is a universal instinct that promotes interest in learning. Play is indirect preparation for future learning.

If something has been playfully learned in one context, it is an act of indirect preparation for learning and applying something new in a different context. New learning happens on the shoulders of the indirect preparation that prior learning provided as playful brain's support meaningful learning.

Play games!

Chess builds the ability to follow a series of logical steps.

Monopoly® demands basic probabilistic reasoning.

Rubik's Cube® fundamentally is an exercise in group theory.

Rush Hour®, a board game, is geometry in the form of a card game.

DragonBox® is an app that helps to learn algebra.

Games show how the spirit of play can be a part of learning. Carol Dweck, who studies the nature of learning, suggested that effort should be emphasized over native ability. This attitude sounds great as an idea but often fails in practice. As an example, a surgeon who succeeds with little effort or a wannabe surgeon who tries mightily, but does not have the eye-hand coordination to succeed. Which surgeon would you prefer? We cannot teach students to do things; we can only teach them to practice things. You coach them in the direction of improving. The sad reality is that not everyone can achieve the level of success required in all fields of endeavor.

"The worst form of inequality is to try to make unequal things equal." (Aristotle)

The personal perceptions that instructors and students have about learning, emotions, and the culture of approaches to learning will influence whether meaningful learning is even possible. Many instructors have been referred to as good communicators.

What is the value of what is being shared? How is what is known being shared?

Poor communication is a problem causing discord between people and countries; businesses; parents and children; teachers and students; employers and employees. Where should we start to find solutions for any miscommunication?

Brain-compatible/student-centered approaches to learning are very simple; to help individuals enhance their ability to learn anything. I would suggest that in learning environments, the design of the approaches to learning be addressed first, followed by the subject of choice. The nature of learning has an unconscious mix of challenges and struggles that are wrapped in curiosity, based on experiences, and prior knowledge. Struggling is paramount when learning (as pointed out by Dr. Robert Bjork).

We can do things for hours without realizing how much time we have spent. Is it motivation or curiosity that leads to losing one's self in a task? What's on the next page? Curiosity leads to motivation. Keep students curious and they will stay motivated. What is this or that? How can I do this or that? In my view, it is curiosity that fuels motivation and drives learning.

Touched by shades of time, the subtle tones of experiences are unconsciously reborn when joined with insights that are coming from an interest in what is going on in the now.

The brain operates by consciously and unconsciously gathering elements and reorganizing them into a cohesive whole. It could be said that the brain is always taking a class and making notes about the patterns and sequences of information that make sense and, based on experiences, have meaning for future reference. The brain's classroom is the environment where life's interactions take place.

When the language of learning is brain-compatible/student-centered, it often has long breaks between personal insights that arrive from the inventory of information and personal perceptions encoded in the unconscious mind. There are those times when mentally walking away; not focusing or giving the brain some breathing room brings on a new or different view enveloped in a rush of adrenaline release. This opens a pathway for developing meaningful learning that is personal in nature.

> **Meaningful learning is dragged down when students hear about what is wrong and about demands that what's wrong needs to be fixed. Freedom from a prison of judgment and criticism that some approaches to learning provide is the aim of information-delivery systems.**

Albert Einstein once said, "Few people are capable of experiencing opinions which differ from the prejudices of their own social environment. Most people are not capable of forming such opinions. Problems we face cannot be solved at the same level of thinking we had when we created them."

TO FOCUS OR NOT?

Read this section with two realities in mind; 1) experiences and prior knowledge influence what we can learn, perform, and create; and 2) information from many different locations in the brain will unconsciously come together to help solve problems.

> Research by neuroscientists from Harvard University and the University of Toronto found that acts of focusing could damage learning and performing. Decisions not to focus or a decreased ability to focus (Attention Deficit Hyperactive Disorder, ADHD), help to ensure a richer mixture of creative thoughts than focusing can produce. The issue, of course, is whether the increase in creative thoughts can be acted upon, or do they simply flash by.
>
> Martha Farah, Ph.D., University of Pennsylvania neuroscientist, "People assume that increased focus is always better, but they do not realize that intense focus comes with real trade-offs, and any big insights are not going to happen." When individuals are not filtering out the world by trying to focus, they end up unconsciously letting in useful information. This occurs without a focused, predictable perspective, as the brain unconsciously considers all sorts of analogies that provide useful insights for learning and solving problems.
>
> "Without focusing, creativity remains in contact with information that is constantly streaming to the brain from the environment." (Prof. Jordan Peterson, University of Toronto neuroscientist)
>
> "Creativity is the result of time wasted." (Albert Einstein)
>
> Dr. Marcus Raichle, Washington University, neurologist, "When your brain is supposedly doing nothing (not focusing) it is actually doing a tremendous amount."
>
> MRI studies by Raichle demonstrated that during non-focusing, or what he calls "a default stage," there was an elaborate electrical conversation going on between the front and back parts of the brain. "I knew that there must be a good reason for all this neural activity, I just didn't know what the reason was." Why was the brain so active during non-focusing or daydreaming? This was the question Raichle asked about a lazy mental process and found the brain was very active during this stage.

In our non-focusing stage, the brain has the ability to blend different kinds of skills and concepts that are encoded in different locations in the brain and can notice what would be overlooked during conscious focus. These brain areas normally do not interact directly. It is when daydreaming without focus that these areas of the brain begin to work in close collaboration with each other for making unconscious associations by connecting new experiences with prior experiences.

Raichle's studies show that instead of responding completely to the outside world, during non-focusing, the brain will start to explore its inner database, searching for relationships in a more relaxed fashion. This relaxed mental process often runs parallel with increased activity in the brain's less linear and more creative right hemisphere.

The brain's ability to blend skills and concepts that are not the same, and that are encoded in different locations in the brain, reduces what would be overlooked if one were consciously focusing. There is an advantage of knowing that insights come from our environment. This can make it easier to have insights that support learning and performing.

Suggestion! **Stop trying so hard to focus and let deep insights just arrive from your non-conscious mind.**

In a relaxed state of mind, it is more likely that the brain's attention will look inward; unconsciously connecting dissimilar information. When we are diligently focused, our attention tends to be directed outward, in the direction of the details of the problem we are trying to solve.

This outward attention prevents us from making the kind of connections in the brain that lead to workable insights. It is fair to say that the answers have been there all along; we just do not allow them to come forward when we are trying to focus.

It is during non-focusing that the brain uses "conceptual blending" the ability to make separate concepts and ideas co-exist, which is crucial to learning, performing, and creativity. Instead of keeping our experiences and ideas separate in our non-conscious mind, the brain will blend them together without our awareness.

The creative powers of the brain and mind amount to little more than the facility of compounding, transposing, augmenting, or diminishing information from our senses and experiences. The brain gathers dissimilar information from many different ideas to support what we are doing in the present. Breakthroughs often arrive when old ideas or past solutions are applied to new situations. Instead of keeping concepts separate, the brain will unconsciously blend them together when we are not trying to focus.

Joydeep Bhattacharya, University of London psychologist, pointed out that interrupting focus brings forward the quiet, unconscious information in the back of our heads that helps learning and creativity. While monitoring brain activity by recording the EEG (electroencephalogram), Bhattacharya and other researchers recognized creative insights before the individuals were consciously aware of them.

Focus is not as useful as attention. It has been shown that focusing can slow down creativity, memory, and problem-solving skills. This is why we remember the answer after we leave the test. Pay attention; just let information flow in or out of the brain. Trying to focus is not a component of the nature of learning, attention is. Rehearsal of just-learned information will help to consolidate it. Rehearsal plus focus enhances retention. Often, students will read a textbook as if it were a novel and expect to retain the relevant information. Students need to read, stop, reread, rehearse, self-test, and explain to others in order to increase retention.

The corporate history of innovation at Minnesota Mining and Manufacturing Company (3M) labs is based on their scientists taking breaks from thinking about ideas for new products. Today 3M sells more than 55,000 products, nearly one product for every employee.

The first essential feature of 3M innovation is their "flexible attention policy," according to Larry Wendling, vice-president in charge of corporate research. Instead of insisting on constant concentration there is a 15% rule. Every researcher spends 15% of their day daydreaming, allowing speculative insights to surface. The only requirement is that the researchers share their ideas with their colleagues. This is how the brain blends different concepts together. The science of insights supports 3M's flexible attention policy.

Other interesting insights into how slowing down or discontinuing conscious focus enhances learning comes from research by Dr. Mark Beeman and Dr. John Kounios, of Drexel University Cognitive Neuroscience Program. They saw a sharp drop in activity in the visual cortex when the brain was paying attention to its own pre-recorded ideas and associations. (Everything about learning, performing, and creativity is supported by associations–if this, then that.)

Beeman pointed out that people often cover or close their eyes, reducing input to the visual cortex, when thinking about solving a problem. This is a major point in the book *Thinking Fast and Slow*, by Daniel Kahneman. Everything we do takes energy, and there is only so much to go around at any one time. One does not engage in a strenuous physical activity and a mental activity, or two mental activities, at the same time. For example, rehearsing an important speech while driving down the freeway, or while doing aerobic exercise on a treadmill.

When the outside world becomes filled with details, the brain will automatically block them out. Messages with few or no distracting details are compatible with the nature of learning. The operative word here is *distraction*. Studying any topic with a radio, iPod®, TV, conversation, or someone tapping in the background provides sensory input that is often difficult to ignore. Focus and concentration may help here for some individuals, but it takes energy to filter out the unworkable input, and this interferes with learning.

Research by Beeman and Kounios demonstrated that people who score high on a standard measure of happiness test solve about 25% more insight puzzles than people who are upset or feeling angry. The relaxed feelings of delight can lead to dramatic increases in creativity and learning. This is hardly surprising considering that one under stress is more likely concerned with survival than some standard test in a lab setting. Recall the previous discussion of stress hormones and the role of motivation in performance.

> **The brain blends concepts that are filed away in different areas to form new connections that we call new insights. Without focusing, the brain starts to explore its inner database, looking for connections in a relaxed fashion.**

AWARENESS

Awareness is something that we have, not something we have to develop. Awareness is a mental capacity that is innate but can be ignored in favor of over thinking. While acts of thinking can seem useful, with the hope of improving critical thinking, the state of natural awareness is what helps to influence our thoughts while expanding their range.

Trying to focus can cause over thinking about results and block natural awareness. Trying to focus is often driven by a culture that is concerned with getting something right. A suspension of an impulse to categorize and evaluate every aspect of our experience in terms of liking and disliking it lets the power of natural awareness expand our relationship with our environment and our responses to it. What better place to begin to tap into the dimension of intelligence found in awareness than in learning environments?

Some believe there is an overwhelming emphasis on information and facts, an approach that can be sorely lacking in awareness is misguided and turns off learning. Overlooked or ignored, the inner life awareness needs to be recognized, attended to, nurtured, developed, and connected with our environment to optimize learning.

Awareness develops the tools to know and recognizes practical opportunities for creativity and imagination. This cultivation begins with exploring the power of being mentally still and what that feels like while being maximally present.

AWARENESS

Alert **W**ith **A**ll **R**esources **E**ngaged **N**aturally **E**asily **S**ensing **S**urroundings

THE BRAIN'S LEARNING SYSTEM

"An understanding of how the brain learns can influence the design of more effective learning." (Center for Education Research and Innovation)

Students will have an interest in learning when the topic matters to them. There is a good chance the topic will matter if the approach to learning creates curiosity. This can be accomplished when the path to learning is wide, not narrow, a key to experiencing meaningful learning.

Approaches to learning should be adventures with no end point, and free of frustration and intimidation, the two main enemies of learning.

Dr. Zeng, of The Allen Institute for Brain Science, found that there are neuronal connections within 295 different regions within the brain. "The direction of the flow of information and the intensity of the connections between regions varies. The strength of some connections was a million times stronger than others, with a small number of very strong connections and a sea of weak connections." The variety in the strength of these connections is influenced by the approach to learning that is being used. (New York Times, Science Section, 4/24/14)

In 1994, fifty-two scholars formulated a scientific consensus and defined "intelligence" as; the ability to reason, plan, solve problems, think abstractly, comprehend complex ideas, learn quickly, and learn from experience; the ability to catch-on, make sense of things, and figure out what to do.

The human brain is not a notebook where information is written down; it is a malleable, living organ that can intuitively assemble proper or improper brain connections depending on how things are being presented to be learned.

> **Effective approaches to learning draw individuals into acts of playful curiosity about their own questions.**

Efficient approaches to learning often ask students to tinker.

Effective approaches to learning do not ask us to merely follow directions, which may not fully engage the higher cortex where learning occurs. They promote self-discovery.

What follows is adapted from *Making Connections: Teaching and the Human Brain*, by Caine and Caine:

> Every brain is unique. Just as there are no two people exactly alike, no two brains are exactly alike. Just as our fingerprints and personalities are different, so are our brains.

Enrichment and novelty are essential to brain growth. Each time you learn or do something new, your neurons make new connections. Our brains like things that are novel. Novelty helps to keep things interesting!

Emotions are critical to learning. During a new learning experience, emotions influence our recall of information and use of skills.

The brain searches for patterns and meaning. When we learn something new, we need to hook that learning onto something we already know.

The mind/body connection has an impact on learning. Learning can be greatly influenced by how we feel physically.

We have social brains. Learning is a social event. We learn by watching, listening, and talking to others.

High stress or threats have a negative impact on learning. Too much stress can damage your brain cells.

We have different brain pathways for different memories; color, movement, rhythm, etc., simultaneously influence different areas of the brain.

The brain is complex and adaptive.

The brain adapts to different environments.

Our brains go through various developmental stages.

There are windows of opportunity, or critical periods, for learning different skills.

The Brain Loves to Play, and Here is Why! presented in a blog on "*AAA State of Play,*" 10/30/14

The brain loves to play! Just like it performs awesome magic tricks as you sleep and age-defying somersaults when you exercise, the brain blossoms during play. Recreation serves our brains in a juicy variety of ways: for children, it is key for neurological development; for adults, it eases stress and anxiety when we play to learn. I think it is safe to say that playtime is one of life's most fun, exciting choices of brain food!

Let's dive a bit deeper into why a child's budding brain needs to play freely and frequently. We all want our children to learn. If school is a delicious buffet of knowledge, recess is the sitting area where kids can enjoy, digest, reflect, and engage upon what they have learned. Active play energizes the cognitive mind and optimizes information absorption. On top of the excellent benefits of play, physical activity alone does wonders for a child's brain and body health! - Also for adults.

No matter how old you are, who does not love to play? While providing more playful opportunities to our children, we should join them to boost our own brainpower! Here are seven reasons why the brain is built and nourished by play:

Creativity Ensues: LEGO comes from the Danish words "leg godt." This translates to "play well." It is fitting that one of the most creative toys has such a clever name. That is because play supercharges creativity. It invites children and adults to explore new connections and discover the power of their hands. Ideas spring naturally from a brain that is fluid and free, thanks to play.

Keep on Learning: Play is a brain's absolute favorite way to learn. It is how we have been learning all along: experimentation, mixing, tinkering, adventuring, and imagining. A child or adult at play is also learning about themselves. Trial and error in the safe, soothing environment of play reveal strengths, weaknesses, cognitive styles, and more.

Challenge Yourself: A playful attitude can make challenges seem less overwhelming and more fun. For example, imagine a group of children pretending to be firefighters. Their objective is to slide UP the playground pole because the stairs back to home base are out of order. It is no easy feat, but the pretend play scenario fuels their dedication to the activity. They communicate, brainstorm ideas, and work together, all while having a merry time!

Relieve Stress: Chronic stress is bad news for your brain. All humans are affected by it. Play is an essential reprieve to the bombardment of information we receive daily. Not only does it help us process all the input, but play also encourages us to uncoil emotions that have bottled up. This is especially crucial for kids as they become familiar with healthy ways to express themselves. Overall, play waters and weeds the garden of our minds.

Improve Relationships with Others: Confidence and communication are big building blocks for healthy relationships. In the pretend play firemen scenario, those children were bouncing ideas around with each other; working as a team to surmount the pole. As adults play a board game, they learn about each other in a lively, stress-free setting. In solo play, we find our inner voice and nourish talents we are proud of.

Learn About Teamwork: Researchers have discovered a connection between frequent cooperation and neuron complexity. With that in mind, imagine a spirited sports game, an intense volleyball tournament, or those firefighters on the playground again. They are all at play, cooperating heartily. Play helps their brains become more open, pliant, and adaptive.

Play is personal and unique to every individual, so I encourage you and your children to find their flavors!

RECAP

Normally a learning model that is general in nature is more effective for retention of skills than a specific expert model.

Some factors that fragment learning include stress, rigid systems, drills, rote learning, expert models, lack of core knowledge, little or no creative play, little sensory stimulation, little opportunity for developing imagination, few chances to solve problems and training in a get it (or did-not-get-it) environment.

Preparation and struggle are two of the most critical stages of long-term progress.

Prior knowledge and experience influence what we can learn, perform, and create; and information from many different locations in the brain will unconsciously come together to help solve problems.

When used as a developmental tool, the process of "playing" becomes the means by which students grow. The playful process provides an opportunity for growth as students learn they can overcome a challenge through the various efforts of fumbling, generating some ineffective alternatives, self-evaluation, and making adjustments.

The state of wide natural awareness is a process that helps to influence our thoughts while expanding their range.

An understanding of how the brain organizes information and learns can influence the design of more effective approaches to learning.

YOUR NOTES ____
____
____

CHAPTER SEVEN

Learn, Develop, Grow

vs.

Teaching-Fixing

When you try to teach, you can win or lose, but when you help individuals learn and invent you always win.

The view here is that teaching-fixing approaches to learning are not only different from student-centered, learn, develop, grow approaches, but they also provide a lower return on investment of time and resources. The short buzz that teaching-fixing environments create is no substitute for the experience that occurs when working with the wonders of the brain's connection to learning. The brain is encoded with emotions, memories, reflections, dreams, and reason that support doing learning.

What are referred to as "grade or outcome learning environments " look for what is right or wrong, not for what is being learned. They are teaching and fixing to get something right.

TEACHING-FIXING to get it right environments tend to be about a student's present skill level. Whereas learn, develop, grow approaches, however, are more about where students are going, and the ultimate outcome can be thrilling.

LEARN, DEVELOP, GROW environments have a sense of purpose that moves beyond trying to fix unworkable outcomes. This happens when the culture of seeing students as apprentices runs deep. Apprentice, one who is learning from practical experience, is from the Latin "apprendere," to learn.

In LEARN, DEVELOP, GROW environments, students are learners in training, moving forward after observing outcomes and seeing both workable and unworkable outcomes as valuable feedback for future reference. "Jumping from failure to failure with undying enthusiasm is the secret of success." (Savas Dimopoulos, Physicist, *New York Times*, 2/25/2014)

In TEACHING-FIXING to get it right environments, unfortunately, students have concerns before they act.

LEARN, DEVELOP, GROW environments motivate individuals in the direction of higher-order thinking, in emotionally safe conditions that support the value of workable and unworkable outcomes.

TEACHING-FIXING to get it right environments often create the kind of stress that releases the kinds of chemicals into the brain that suppress learning.

In LEARN, DEVELOP, GROW environments curiosity may be the most useful way of looking at motivation and the nature of learning. Keep students curious, and they stay motivated, which is compatible with the nature of learning. TEACHING-FIXING to get it right environments often misjudge the needs of their audience. In those environments, the teacher often lacks the insights that support meaningful learning.

LEARN, DEVELOP, GROW approaches are interested more in the way people are thinking creatively, than any outcome from following directions. Student-centered/brain-compatible approaches to improving develop agile minds, puzzle solvers capable of thinking effectively, on the spot, in ever-changing conditions.

LEARN, DEVELOP, GROW information-delivery systems are supported by a culture that is benevolent, tolerant, understanding, flexible, and positive about unworkable outcomes. TEACHING-FIXING approaches to learning are based on a culture that views poor outcomes as a failure and not as the valuable feedback that nature intended them to be.

LEARN, DEVELOP, GROW approaches to learning have nothing fancy, complicated, or frustrating about them. Within this culture, students are given the opportunity to rethink their relationship with their trial-and-error risks that are clearly important components of change and development. The value of inconsistency is embraced in LEARN, DEVELOP, GROW environments.

TEACHING-FIXING to get it right approaches, unfortunately, tend to see proficiency as a fixed, right, or wrong external quality. On the other hand, student-centered LEARN, DEVELOP, GROW environments recognize the internal personal qualities of originality, problem-solving, open minds, and creativity as intellectual tools. These environments integrate the brain's internal thinking skills for gaining an education.

TEACHING-FIXING to get it right approaches often give the illusion of progress, without the rewards of long-term learning.

TEACHING-FIXING approaches have students trying to learn consciously, which can be accomplished non-consciously, through playful student-centered/brain-compatible approaches, free of frustration and intimidation.

In LEARN, DEVELOP, GROW environments, the flow of unjudged actions is never-ending going in the direction of reaching full potential.

LEARN, DEVELOP, GROW approaches use strategies that support independence, curiosity, drive, acceptance of all outcomes while allowing outcomes to arrive. "The road to discovery is not found in the intellect. There comes a leap in consciousness, call it intuition or what you will, and the solution comes to you, and you do not know how." (Albert Einstein)

TEACHING-FIXING to get it right environments often use an approach that can unknowingly filibuster opportunities to gain meaningful learning. How do students go from not knowing to knowing? What becomes the first light? The answer here suggests, join the art of teaching with the science of learning. "Not everything that can be counted - counts, and not everything that counts can be counted." (Albert Einstein)

> **The knack of rethinking; the knack of seeing more options; the knack of developing other usages for information; are all born in learn-and-develop environments free of judgments while promoting self-organization, self-development, and self-evaluation.**

A suffocating approach to learning often exists in teaching-fixing environments filled with emotionally damaging criticism. On the other hand, student-centered/brain-compatible learning-developing environments are organized to serve students, not to legislate to or judge them. It is an approach that submerges its voice beneath that of the student's personal observations and experiences. In student-centered/brain-compatible approaches to learning, the message to students is to absorb and allow, with students remaking themselves, often without realizing change is taking place.

The kind of stress that slows learning normally exists in a teaching-fixing to get it right environment. I use the terms "teaching-fixing to get it right" vs "learn, develop, grow." Professor Terry Doyle uses the terms "teacher-centered environment" vs "learner-centered environments" in his book *Helping Students Learn in a Learner-Centered Environment.*

Doyle states, "Our desire is to incorporate the new discoveries about how the human brain learns; teaching in harmony with the human brain." Learn, develop, grow environments have jumping off points that are not available in teaching-fixing approaches. Efficient acts of learning often do not rely on rational or logical approaches to insights and answers on which teaching-fixing environments are often founded.

> ***Suggestion,*** **when it comes to learning, do not strive for perfection, embrace natural evolution. "Evolution is a process of gradual changes between generations that were not planned, designed, or deliberate; they just happened and will occur in any living organism which involves copying information." (Al-Chalabi, et al., 2008; 26) The same holds true for most new learning; it was not planned or deliberate but did develop and was encoded within our brain.**

The Committee on Learning Research and Educational Practices in America stated, "learning-assisted environments that guide, motivate, and support self-discovery learning are more effective than teacher-centered environments. Learning skills and information do not come from teachers." Research has found that interactive approaches to learning are better than a traditional lecture approach.

STUDENTS IN TEACHING-FIXING

Students can participate in their own lack of progress when they see value in teaching-fixing environments and the directions and corrections they provide. When all is said and done, there is often more being said than being learned by students in teaching-fixing environments, with the lack of meaningful learning going unrecognized.

> Some approaches to learning are little more than a reaction to poor outcomes and do not provide proactive, emotionally positive experiences. A proactive experience develops the self-confidence and self-assessment skills that support experiencing meaningful learning. This important connection is often overlooked, but it accounts for how different teaching-fixing it get it right environments are from learn, develop, grow environments. The brain's ability to pay attention is very impressive under the right circumstances!

Teaching-Fixing	*vs*	Learn, Develop, Grow
Provides information	*vs*	Curiosity-driven
Has a specific goal	*vs*	A process that never ends
Asking you to follow	*vs*	Wants you to investigate & invent
See wrong answers	*vs*	Sees feedback for future reference
Challenges our emotions	*vs*	Protects and guides our emotions
Can cause frustration	*vs*	Is a safe environment

FIXING IS ATTRACTIVE

Fixing is attractive to the emotional needs of a student's mind. Hopefully, the emotional mind will learn that there is no failure, only usable feedback for future reference. Progress and change from learn, develop, grow approaches to education tend to stay around and are not just a novelty of occasional success. Unfortunately, teaching-fixing environments can foster a damaging fixing mentality, thereby prompting students to attempt perfection that can last a lifetime.

There is a need for what could be called a "wonder approach" to learning that uses what-if thinking connected to the nature of learning. Learn, develop, grow approaches to improving use an autobiographical path to make sense of the world in ways that others may not. Self-organization and adaptive mimicking support the brain's ability to learn, develop and grow while applying core information.

Learn-develop-grow systems help students to make personal sense of the environment. I would describe the brain's learn, develop, and grow operating principles by saying: human beings of all ages learn by doing, observing the outcome, then adjusting if needed, based on past experiences. Professor Peter Atkins said something similar; "Human nature makes an observation, then compares notes."

At the core of learning and developing is mankind's natural drive, or intent to do, learn, and survive. Research points out that the to-do intent always precedes ability. One purpose of nature's biological intent, or plan for learning, is to indirectly prepare functions that will become fully operative much later.

There are teaching-fixing to get it right environments that suppress mankind's natural intent to do by sidestepping the use of indirect preparation that past experiences provide when how-to directions are given. Some things learned by rote are normally half-understood and imperfectly felt (force-fed information).

Teaching-fixing environments often mute curiosity and dilute imagination with passive, non-invasive approaches to learning. On the other hand, efficient student-centered learn-and-develop environments are safe and active, as individuals learn more from ever-changing environments by using real-time interactions with basic information.

Student-centered learn-and-develop approaches pull the indirect preparation of experiences forward; mixing them with what is currently taking place. Efficient use of our experiences brings out autobiographical images and personal visions that support the kind of meaningful learning that lasts.

Learn-develop-grow environments support using one's personal genius, uncovering the creative potential of individuals that teaching-fixing environments do not promote. Individuals often find themselves caught in the haze of how-to directions from a perceived expert and not at the hub of self-organized creative insights that can guide a learn-develop-grow journey in the direction of progress.

I realize that it will take time to change the powerful assumptions about teaching-fixing environments that have experts delivering how-to statements to individuals who want to make progress. A learn, develop, grow approach to education is shaping a student's interest and enthusiasm for seeing options that enhance their ability to ask useful questions.

This process of learning also recognizes that silent personal insights are very valuable for sustained progress. In learn-develop-grow environments, growth is just ahead, with solutions from the adaptive skills of curiosity, imagination and what-if ideas unleashing the student's creative potential.

Plato reasoned that the most important knowledge was already inside the student. Therefore, the role of the teacher was to facilitate the student's realization of this inner knowing through a quest that leads to independent thought. I refer to this realization of inner knowledge as "out-struction"– drawing out based on individual experiences–as opposed to instruction.

Efficient approaches to learning help individuals invent and create their own reasoning and problem-solving skills. Meaningful acts of learning rarely rely upon the completely rational or linear approach to answers on which teaching-fixing environments are normally founded.

Learn-develop-grow approaches to progress allow our current observations to interact creatively with our personal experiences. Studies show that the nature of learning uses cross-disciplinary reasoning and deduction skills, based on personal experiences that can be overlooked in teaching-fixing environments.

Teaching-fixing approaches to progress normally do not give students the opportunity to make up their own minds. The opportunity to choose may be the most important element to gaining a good education. How-to directions in teaching-fixing environments can reach such abstraction that valuable core subject knowledge gets lost, thereby suppressing a huge number of useful personal questions and insights.

On the other hand, learn-develop-grow environments have the ability to illuminate the kind of reference points and associations that make what may seem difficult, easy. Learn-develop-grow environments expand established patterns and create new personal pathways for putting together new information.

Learn-develop-grow approaches to education provide a quality learning experience by focusing on three-dimensional integrated learning (mental intelligence, physical intelligence, and emotional intelligence) for enhanced learning ability. By staying aware of these three areas of intelligence and the brain's five major learning systems that UCLA's Learning and Forgetting Lab point to (Reflection, Action, Intention, Social, and Passion) one can gain valuable insights into the nature of learning and long-term progress.

> **In learn, develop, growth environments students are active participants in their own learning. They are sharing their information, and they are practicing their learning with the educator.**

If learning is not personally meaningful, the information will not be remembered. The age-old question, "How are you going to use this?" must be answered and this is often the forgotten part of approaches to learning.

POOR ASSUMPTIONS

In their book, *Education on the Edge of Possibility*, Renate and Geoffrey Caine point out some new ways of evaluating approaches to learning and teaching. Their insights are founded on research about the brain, questioning many of the public's and some educator's deepest assumptions about education.

These poor assumptions, for the most part, are based on teaching-fixing approaches to learning that see educators and perceived experts as owning information that they then pass on. These teaching-fixing environments are not as efficient for encoding for long-term learning (retention). In teaching-fixing environments, individuals are asked to focus on getting it right or on how to do something and subsequently avoid self-discovery experiences.

On the other hand, an inside-to-outside approach to learning develops insights about core concepts. These approaches move away from details and move on to basic core information allowing any details that follow to be in accordance with one's own individual make-up. Contemplate this: Core concepts are based on the environment.

We shoot basketballs up to the hoop because the basket is up. Any details of shooting the ball up follow that basic core concept. When details lead, as they do in teaching-fixing environments, it can take a long time to learn or learning may never take hold.

In learn, develop, grow environments, individuals are always given the freedom to self-organize and learn through being open to all possibilities. This is their opportunity to make their mind up for themselves. The Caines point out that these approaches to learning develop individuals who are self-confident and can be in charge when left on their own. It seems these self-organizers are happy with themselves and their skills. Their creative and innovative thinking skills put them in great demand in the workplace.

Learning does not develop from scratch. Comparisons, associations, and parallels promote learning. When first learning a basic skill, the foundation for learning that skill is formed when the brain sees parallel relationships to our retained experiences. Long-term learning is normally not formed through memorization, doing drills, or copying expert models.

Hopefully, as individuals encounter new insights, it will lead them to the kind of curiosity and use of imagination that develop know-how skills that are personal in nature. Change, transition, or modification occurs at a lasting level when attempts to make progress are cross-referenced with our experiences that have been self-organized in learn-develop-grow environments. As cognitive scientists say, the more parallels and association, the merrier.

Efficient learn, develop, grow environments are less concerned with outcome results than the development of the performing self. Learning starts to take hold when we master nature's most basic opportunity for learning, which is just do something. As sure as night is normally dark and the day is normally light, learning beckons the law of chance. Without chance and doing, there are no new beginnings.

For thousands of years, when mankind woke up in the morning, our ancestors' brains were wired to look for food for their bodies, an activity that was food for their minds. If food for the body had always been given to man, mankind would not have self-developed this ability to find food, and we would not have survived.

Likewise, if our ancestors did not have the opportunity to self-develop the steps and stages of solving problems (food for the mind) we would not have survived, but humankind has survived through the development of the self. Learn, develop, grow environments recapture one's childhood where acts of curiosity lead to learning that lasts.

When educators use words, symbols, or phrases for their own purpose, they often miss the student's needs. The structure and design of learning environments that

are expected to be efficient must be influenced by the needs of the individuals they serve.

The core of meaningful learning is more about improvising and inventing than following directions. The intention of teaching-fixing environments normally promotes ideas overlooking the value of self-discovery by students as accommodations to the task at hand.

John Taylor Gatto, the award-winning educator, stated in his book, *Dumbing Us Down*, "The truth is that some schools don't really teach anything except how to obey orders. We've built a way of life that depends on people doing what they are told because they don't know how to tell themselves what to do. That's the biggest lesson I teach."

RECAP

What are referred to as "grade or outcome learning environments" look for what is right or wrong, not for what is being learned. That environment is a teaching and fixing to get something right.

A suffocating approach to learning often exists in teaching-fixing environments that tend to be filled with emotionally damaging criticism.

Student-centered/brain-compatible learning-developing environments are organized to serve students, not to legislate to or judge them.

Learn-develop-grow approaches to change and improving use an autobiographical personal process to make sense of the world in ways that others may not.

Learning does not develop from scratch. Comparisons, associations, and parallels promote learning. When initially learning a basic skill, the foundation for learning that skill is formed when the brain sees parallel relationships in our experiences. Long-term learning is often not formed through memorization, doing drills, or copying expert models.

Efficient learn-develop-grow environments are less concerned with outcome results than the development of the performing self. Learning starts to take hold when we use nature's most basic opportunity for learning, which is *just do something*.

YOUR NOTES ______________________________

CHAPTER EIGHT

Approaches To Learning Are Different Than Approaches To Teaching

Information has always been an important resource in our life, and how it is delivered is critical. Skill storage, continuous learning, performance management, talent mobility, and long-term retention are all available to students when methods used in the approach to learning leverage the brain's connection to learning.

Learning that is organized around clear questions, motivated by the student's curiosity, encourages them to wonder, gain insights, analyze information, make changes in their thinking and behavior, to achieve new understanding. In this environment, students reason, think critically and apply their knowledge in meaningful ways to make connections between content and their expectations. If the nature of learning could speak, it might say, "Train Your Brain," which is also the title of a book by Dr. Ryuta Kawashima.

> **Mark Twain said, "The two most important days of your life are the day you are born and the day you realize why." We were all born to learn, develop, and grow. Do some approaches to learning fail the students before the students themselves fail during learning opportunities?**

Being fit is usually associated with the physical body, but approaches to learning can influence what some call "cognitive fitness" and our problem-solving skills positively or negatively. "Fitness measures the ability to perform and adapt as we encounter problems and disturbances." (MIND CODE, by Charles Bailey, founder of the Global Institute for Scientific Thinking; 7) This holds true for physical and mental abilities.

When attending the 2015 Learning and Brain Conference, it was mentioned that brain imaging is a more useful predictor of learning than behavioral evaluation. This could not have been said ten years ago. The use of functional magnetic resonance imaging (fMRI) has revolutionized the study of the brain in action.

Over time, several learning theories have proved useful including:

1985, Bruner's cognitivism

1950, Piaget's constructivism

1998, Orkwis & McLane's universal design for learning

1991, Lave & Wenger's situated learning

1999, Tomlinson's differentiation

2001, Revised Bloom's Taxonomy by Anderson & Krathwohl

2005, Webb's depth of knowledge

Note: to be effective all these approaches recognized that the BRAIN is not merely a black box that organizes and distills information.

ENHANCE LEARNING WITH QUESTIONS

Questions that can enhance the work of the brain's higher order thinking and approaches meaningful learning:

What do you think caused that outcome?

What may happen next?

What has been left out?

How can this be modified?

What is the most important element?

What did you first notice?

How many ways can this be done?

What would be a metaphor for this?

What do you want to do next?

"Once you organize your mind -you will be able to find what you want quickly."
Tamora Pierce, *Wild Magic*.

The learner's brain wires itself in reaction to the learning environment it is experiencing. Is that environment supportive or unsupportive, motivational or suppressive, safe or unsafe, judgmental or proactive, developmental or corrective?

THERE IS A DIFFERENCE

What kind of messages should approaches to learning send to students? Keep the old saying "quality vs quantity" in mind when designing approaches to learning, teaching, and performing. Some approaches to learning are filled with information that I call "intellectually interesting but educationally vacant" because of the way it is being delivered and the amount of details and data that were given.

Benedict Carey, author of *How We Learn*, pointed out that "misconceptions about learning are based on an imperfect understanding of people as learners." Dr. Robert Bjork points out, "How we learn is different from how we think we learn."

Learning how to learn has been called the most important skill of our time by leading cognitive researchers and neuroscientists. Perhaps the best way to approach learning is to have some understanding of what suppresses and what supports learning. Student-centered/brain-compatible approaches to learning-development tell the difference between:

Instruction that is In Context	**Out of Context Instruction**
Coaching	Teaching
Learning	Fixing
Deep Learning	Wide learning
Learning and Developing	Trying to get right
Creating	Following
Feedback	Failure
Self-Discovery	Following Directions
Proactive Approaches	Just Reacting to outcomes
Different	Better
Making Connections	No Connection
Intrinsic Learning	Extrinsic Learning
Mindful	Mindless
Informed	Uninformed
Unconscious Learning	Conscious Learning
Randomness	Perfection
Acts of Learning	Acts of Teaching
Applying information	Just Knowing
General Concepts	Specific Information
Learning Models	Expert Models
Emotionally and Physically Safe	Unsafe

INSIDER INFORMATION

There is a back story to gaining an education that is often overlooked. It should be called an "inside story" for two reasons: first, the brain's internal operations and next, the information that is already encoded in the brain. Both reasons are linked to the nature of learning anything.

What is referred to as "insider information" in the investment industry can be a metaphor for the inside story of learning. Insider information in the investment industry is illegal because it provides a big advantage to those have it. Now play with the idea that when it comes to acts of learning, every human being has the use of insider information in the form of prior knowledge and the internal operations of the brain that provide support for experiencing meaningful learning.

The existence of this insider information reveals that organizing approaches to learning with the brain in mind is an important first step. Student-centered /brain compatible approaches to learning are designed to develop and rely on "mindsets before skill sets," or insider information. When individuals experience anything, THE BRAIN WAS INVOLVED FIRST, often below our awareness.

OUT OF YOUR MIND LEARNING

The phrase "out of your mind" normally has negative connotations, but when it comes to learning, out of your mind describes where meaningful learning comes from. For example, how the inner voice of the mind influences acts of learning was an insight used by the great NFL coach, Jimmy Johnson. He was known for telling his football team, "Your mind controls your body." This insight was also promoted by educational researcher Eric Jensen, who organized the popular and frequently sold out workshops, and at Harvard's Graduate School of Education (home to the Connecting the Mind - Brain to Education Institute).

Today, I ask students if you had a choice what door would you go through? Door "A" that reads "Enter if you want to be provided with someone else's knowledge," or door "B" that reads "Enter if you want to learn"? Every individual that I have asked those questions replied that they would go through door "B" where they were going to learn.

That answer shows me that a student, who may be uninformed about the brain's connection to learning, has a natural instinct telling them what studies compiled here also say. Teaching-fixing environments are not only different from learning-developing-growth environments; they give a lower return on the investment of time and resources than brain-compatible/student-centered learning environments.

The statements that learning is fundamental to human beings; learning is the specialization that we use to become fully human; learning is a process of change that is mostly unconscious, below our awareness, are pointed out in the paper, The Fundamental Importance of *The Brain and Learning for Education in 2008*, co-authored by Kurt W. Fischer (Harvard University) and Mary Helen Immordino-Yang (University of South California).

When the information is delivered with the brain in mind, processing information, situation adaptability, abstract thinking, creativity, visual recognition, organizing skills, fluid intelligence, flexible thinking, skill storage, emotional stability, continuous learning, performance management, talent mobility, reasoning and deduction skills, imagination and long-term retention are all available.

Brains do not operate in isolation–they need context. Context includes:

> **A Physical Context: external world, internal tools**
>
> **A Psychological Context: historical context, cultural setting, social context, emotional experiences (feelings, beliefs, goals, values)**
>
> **A Narrative Context: Logical/quantitative, existential, hands on, interpersonal collaboration, aesthetic, words.**

INDIRECT PREPARATION SUPPORTS LEARNING

With efficient preparation, our ability to learn over time takes hold. It may help to see this preparation as having two stages. One stage occurred in the environment during evolution and natural selection. The other stage of preparation is driven by man's urges and instincts to do and be in the know.

This to-do instinct encourages people to be curious and develop a variety of insights that prepare them for learning anything. Before man could emulate, he had to be prepared. One of the most valuable insights about the nature of learning is that learning always takes place against a backdrop of existing knowledge or preparation that was mostly indirect. Involvement with interactions in one area of the environment is "indirect preparation" for creating workable interactions in dissimilar environments.

When we are doing anything, we may not recognize that this act of doing is based on a record of prior knowledge, and at the same time is adding to an inventory of information that can guide future acts, both similar and dissimilar. All elements of our interaction, physical and mental, conscious and non-conscious, with our environment, become indirect preparation for the next interactions, both similar and dissimilar. Indirect preparation has a broad range of outcomes, some workable, some similar, some unworkable, that develop flexible knowledge and transferable skills. This is one of the essential elements of long-term progress, an insight that is often overlooked and undervalued.

One of the messages at the 11th annual Teaching Professionals Conference held in Boston in May 2014 is:

Brain-compatible/student-centered learning is a team event seen as trillions of brain neuron connections that team up with a sense of freedom and a hint of human imperfection.

Approaches to learning that are compatible with the nature of learning deal more with the student's brain, than the subject.

"Techniques Used in Teaching More Important than Who Teacher Is" was the headline of an Associated Press story in May of 2011. The article pointed out how it is the strategy of the approach to learning information and skills that influence the pace of progress that students experience.

Indirect preparation is a huge gift of nature; it is an ongoing learning process that allows us always to be in the process of updating our mental and physical skills. The brain is constantly using past and present, similar and dissimilar, interactions to rewire itself in our ever-changing world to use for future interactions.

Nature's plan for learning is not concerned with perfect models. General concepts will do just fine. For example, once a child or adult is launched on his attempts by urges and instincts to do and to learn, he will often improve on any examples set for him. A learner's to-do instincts are designed to drive him to do something, anything definite. In these acts of doing, people learn about organizing, interacting, and staying on task, while developing insights into what works and what does not.

People enhance the steps and stages of learning every time they do something definite, (perhaps not perfectly), and then they advance beyond any examples that may inspire them to use their urges and instincts. The urge to do something definite that turn out unworkable should not be fixed; they should be observed. The unworkable outcome of the to-do instinct is not a failure; it is valuable feedback for future reference.

> **People organize their preparation skills for future learning through what could be viewed as indirect preparation. One of the most important insights about learning is that it always takes place against a backdrop of existing knowledge. All experiences are indirect preparation for learning something else.**

The term "transfer of learning" refers to when new learning is joined with experience, allowing new information to move from short-term memory to long-term memory. When we are learning new information, depending on one's experience, the information is either set aside as not useful or encoded as useful for future reference. Experience reveals our options when learning, not the answer.

Individuals of all age groups can indirectly enhance their ability to learn through what they experience (similar and dissimilar) beyond formal education settings. This holds true while people are in school, attending business seminars, participating in workshops, or involved in any other learning environment.

In nature's plan for learning, first attempts at learning anything are attempts that develop the skills of preparing to learn something different. This subtle insight is not so subtle when it comes to having accurate insights about the nature and joys of learning. Learning is the result of indirect preparation or what has been previously encoded in our brain and mind.

The influences of indirect preparation on direct learning should not be undervalued.

The idea of exploration, or scouting, should become an element of any approach to education. It is not a good thing to split life into living and learning. Robert Sadoski, Ph.D., Stanford University, points out that memory from indirect preparation leads to new learning, which leads to behavior modification.

Meaningful learning is based on experiencing all the stages (the ups and downs) of preparation. This is nature's indirect approach and preparation for learning. It is a poor concept that if a model (an example) were copied exactly, it would help students to learn to do likewise. Expert models are best used as motivation, drawing people to an activity, not as a blueprint to copy.

NON-CONSCIOUS MIND

Dr. Emanuel Donchin, University of South Florida, "Most of our learning does not even require that we pay conscious attention, 99% of all learning is a non-conscious process." On the other hand, asking a question is an externally-prompted path to learning that is brain-compatible and requires conscious retrieval of knowledge that may have been encoded non-consciously.

"The non-conscious capacity of people to acquire information is much more sophisticated and rapid than their conscious capacity to do this. Humans have no conscious access to the non-conscious process that they use to acquire knowledge. Indeed, research into the unconscious acquisition of knowledge demonstrates that the human being has an enormous capacity non-consciously to make inferences." (J. Weiss, The role of pathogenic beliefs in psychic reality. *Psychoanalytic Psychology*, Vol 14(3), 1997, 427-434)

Weiss points out that there are two levels of the mind, one is non-conscious, the other conscious. The non-conscious can be more important than the conscious for guiding what we do. The non-conscious is filled with the raw material of experiences that influence our reasoning, learning, and decisions. The conscious mind has no record of experiences and can only process less than 50 bits of information per second while the non-conscious mind can process millions of bits in the same timeframe.

David Eagleman, a neuroscientist at Baylor College of Medicine, points out that only 5 -10% of our brain is dedicated to conscious behavior. It is the unconscious workings of the brain that are crucial to everyday functioning including learning. Think about how many things we do every day that would look like we are consciously thinking to accomplish them, or trying to learn from them, but we are not.

The Seabourn Company had an ad in the New York Times that read, "Delight in never having to ask. The art of delight is that it knows what you want before you do."

Our non-conscious mind provides a similar service, or delight, by pulling forward information before we are consciously aware of it. There are comfort and learning to be found in the silence that surrounds the activity in our non-conscious minds. A location that supports experiencing meaningful learning is defined by the nature of being human. It is by being a learner, not a follower that we learn.

"Most cognitive processes have been found to occur non-consciously, with only the end product reaching the conscious mind, if at all. (Joseph E. LeDoux, Professor of Neuroscience and Psychology at New York University) Contrary to popular beliefs, a cognitive process involves a non-conscious processing mechanism." (*Human*, Michael Gazzaniga, 2009; 67)

The mental process involved in learning a skill or information moves from the higher conscious areas of the brain in the frontal lobe to the more primal areas in the back of the brain that control interactions, often executed without conscious thought. We are not wired to fail but to search and be curious. This natural process can be damaged in teaching and fixing to get-it-right approaches to learning.

Students who experience meaningful learning do not simply remember what they are told. They construct their own meaning and understanding based on their personal experiences; a non-conscious process that we are unaware of is taking place. This reality is why stories and metaphors are important in learning environments.

> **Everything is always based on the story we tell our self. Everything I see, hear, taste, feel, or smell, is evaluated unconsciously by prior knowledge.**

In his book, *Incognito*, David Eagleman notes that there is a gap between the amount of information the brain consciously knows and what the non-conscious mind is capable of accessing. We are not aware of the vast majority of the brain's ongoing non-conscious activities, nor would we want to be. If we were, that would interfere with the brain's highly efficient non-conscious capabilities for learning and performing.

The brain secretly performs an enormous amount of work incognito. Eagleman notes that "The best way to mess up when playing the piano; or get out of breath; or miss a golf shot is to consciously analyze what you are doing."

Eagleman points out that there is a difference between knowledge and awareness.

Implicit memory is completely separate from explicit memory. Implicitly learned information (non-consciously) is simply not accessible on command to consciousness. It shows up on an as needed basis. Unconscious learning and most everything about our interactions with the world rests on this reality. *Aha!* moments do not come on command; they just arrive!

In the late 1800s, Hermann Ebbinghaus wrote, "Most experiences remain concealed from consciousness, and yet they produce an effect that is significant, which indicates their previous existence." The best reason we have for believing that unconscious learning exists is that we have all experienced it.

> **The brain operates and learns by making predictions, influenced by information gained from all the differences found in prior outcomes encoded in the non-conscious mind. Random, spontaneous, improvisational thoughts and actions share a common element, *a result.***

These results are all different, creating a variety of useful non-conscious reference points that become the tools of predictions for future use (if this, then that). This kind of spontaneous and improvisational learning can support reaching one's potential with any skill or subject.

Yes, there is an alphabet, and there are the accepted rules of grammar, but rules cannot tell you what to think or how to create what you personally want to do or say. Systems and rules have limits when it comes to creativity. Any insights about the topic of learning should travel beyond the content to be learned and into the silent language of the process of non-conscious learning.

Ultimately, the only information and skills that help one to experience meaningful long-term learning are those invented, experienced or organized personally. What does it look like to you? What does it feel like to you? What did you think about it?

When it comes to learning, the influence of the unseen raw material already stored in a non-conscious mind is similar to an off-duty supervisor having influence over what happens at work even though they are not present. Perhaps we could say there really is no new learning. What we already know and have stored in our non-conscious mind will have influence over what can be learned, similar to the unseen supervisor's power to influence.

Implicit and Explicit Memory

By *implicit instruction*, we refer to teaching where the instructor does not outline such goals or make such explanations overtly, but rather simply presents the information or problem to the student and allows the student to make their own conclusions and create their own conceptual structures and assimilate the information in the way that makes the most sense to them.

By *explicit instruction*, we mean teaching where the instructor clearly outlines what the learning goals are for the student, and offers clear, unambiguous explanations of the skills and information structures they are presenting.

LEARNING FROM DAY ONE

Our progress or lack of progress in learning environments is influenced by the kind of chemicals the approach to learning releases into our brain. Think of it as similar to how some food is said to release good calories and other food offers good taste but release bad calories. For example, when we like something or someone, or when we want to do something again, it is because we want more of the pleasure chemicals that choice releases into our brain. We want more of how that chemical made us feel.

This is similar to how the quality of learning we can experience is based on how the chemicals that the approach to learning releases into the brain makes people feel about themselves, the topic, the teacher, their abilities, and physical and emotional safety when learning.

When learning; "our hormones are important, they are responsible for transporting information throughout our nervous system. The release of some particular chemical hormones increases our capacity to transmit information and support efficient learning. In addition, the release of other chemical hormones will reduce this capacity." (Brandsford, 2007; 163)

Designed by nature and nurture, we could say that the brain was the world's first "resource catalog," with the ability for holding and finding information, perhaps providing a blueprint for Goggle. "The brain is a chemical-electrical powerhouse, sending messages where needed, in a perfectly targeted way." (*The Brain*, Al-Chalabi, Delamont, & Turner, 2008)

"The brain organizes human behavior; any questions about why we behave as we do will have an answer in the circuits of the brain." (Al-Chalabi, et al., 2008; 96)

RETHINKING APPROACHES

The following statements are excerpts from Richard Perez Pena's article about problems in education. (*New York Times*, 12/26/14)

The University of Colorado, a national leader in the overhaul of traditional teaching methods, has been evaluating students' learning over several years. Given the strength of the research findings, it seems that all universities would be desperately trying to get into the act but, unfortunately, they are not. There are many explanations why there is less interest than there should be in enhancing acts of teaching.

“ What drives advancement at universities is publishing research and winning grants. Teaching is not a very high priority.” (Marc T. Facciotti, Associate Professor, University of North Carolina)

> **Noah D. Finkelstein, a Physics Professor and Director of Colorado’s overhaul of teaching efforts said, “Faculty does not like being told what to do, but there are plenty of studies that say what they are doing is wrong.” These studies showed professors that their students were learning less than they thought.**

Of course, telling experienced teachers that they need to learn how to teach does not always go over very well. “There is some ego involved, and it is hard to hear that what you have been doing doesn’t necessarily work,” said Dr. Mitch Singer, the first professor at the University of California Davis campus to teach a new-style class. This method included a brain-compatible/student-centered approach of putting students from large classes into small groups, experimenting, avoiding questions that required yes or no answers, using questions that have two or three follow up queries; as weak spots are revealed, lessons are tailored accordingly.

“It is already night and day.” Student, Jasmine Do “this makes you participate and pay attention because there is always something new going on.” (Dr. Singer)

The overhaul project borrows elements from many sources including the University of California at Davis; more than a decade of work from the University of Colorado and other institutions; software from Open Learning Initiative at Carnegie Mellon University; research by Carl E. Wieman, a Nobel Prize-winning physicist at Stanford who founded Colorado ‘s project and a parallel effort at the University of British Columbia; Eric Mazur, a Harvard physicist and author of the book *Peer Instruction*; and Doug Lemov, a former teacher and author of *Teach Like a Champion*. While some educators at the lower levels receive some training in education theory, most college professors acquire none.

“Teaching is no different from any other human-to-human interaction. If our students do not see the relevance of what it is we want them to learn and how it can be applied in their lives now or in the future, they are less likely to engage. If they do not understand the content, skills, or behaviors being taught and don’t find that learning enjoyable and challenging they will struggle to stay involved. Finally, if they do not feel respected, valued, emotionally and intellectually safe they are much more likely to withdraw rather than commit to the teacher-student relationship which drives learning in and out of the classroom.” (Professor Terry Doyle)

POSITIVE LEARNING

Rosabeth Moss Kanter, a Harvard Business School Professor and author of sixteen books said, “Confidence, a belief that the outcome of our efforts will be positive, is nurtured by others.

Educators, coaches, guides, and students should pay attention to what is good, not what is bad. Look for small wins that will support bigger successes later on and for milestones to move on from."

What I am calling "brain-compatible/student-centered" approaches to learning is written about in a paper titled *Positive Assessment Form of Learning*. The authors, Sara Leist and Kathryn Jensen said, "We have argued that an absence of an investment in the understanding and the refining of positive assessment practices is a notable oversight in the field of education and sports. This significance is both a moral obligation, as well as a valuable condition for ensuring that any assessment decisions are learning oriented." (*International Review of Research in Open and Distance Learning*, 2/1/2111)

Educational practices and outcomes can be enhanced through the establishment of positive assessment practices that are learning oriented. The engagement of students and athletes in the collection, interpretation, and implications of acts of positive assessment can enhance outcomes when learning. Purposeful positive-assessment involves students planning to collect performance evidence for informing ongoing learning and performance when development, not fixing is the aim.

Feedback, in itself, may not promote learning unless students are positively engaged with it. (Gibbs, Simpson) UCLA Learning Lab found that the use of teaching aids often provides too much support, thereby suppressing learning. Regardless of the demand on the educator or coach, positive-assessment learning is central to quality teaching and coaching, even at the high end of the professional performance spectrum.

Learning and pedagogical concepts have traditionally been outside the domain of education and sports. However, contemporary research has demonstrated that coaching is really about learning. "Pedagogy" relates primarily to the interactions between the educator and student, coach, and player. On the other hand, a "learning-message" refers to the information both explicit and implicit that is necessary for performing, including rules, tactics, and skills.

The current use and value of positive assessment components of teaching and coaching are often notably absent. Researchers now believe that this is a significant oversight. Failing to recognize this key aspect of learning and pedagogy overlooks opportunities for optimizing the development of teachers and coaches, students and athletes.

The components of efficient positive assessment include learning orientation, authentic, validity and social assessments. Positive assessments support students by applying the necessary what-to-do, not what to fix, knowledge during learning for themselves.

Learning-orientation

Learning-Oriented assessments provide positive what-to-do information that is accessible and interpretable by learners and instructors alike.

Authentic

Authentic assessment is concerned with the relationship between learning content and the context of the assessment used. Assessments must be positive and meaningful to the learner and relevant to the context in which the knowledge and skills will be used. This involves a demonstration of knowledge and competencies in the mode in which they will be engaged in a performance context.

Validity

Validity assessment must satisfy six conditions: content, substance, structural, generalizable, external, and consequential elements. In other words, what needs to be valid is the information collected, the means of its collection, the interpretation and the positive implications and consequences for learning.

Social

Socially, all learners are given equal opportunities to engage in positive assessment. This includes the importance of providing multiple opportunities to demonstrate evidence of performance in ever-changing environments, ensuring that learners are engaging positive assessment in a context that suggests learning. Learning-oriented positive assessments help people learn to engage in the process and become familiar with the standards they are going after. In these environments, the person learns to understand better and appreciate their own strengths and areas that further their development.

The term "feed-forward" was coined by D. Bond to indicate the notion of using positive-assessment feedback to assist in the improvement of future performances, allowing athletes to become assessors of their own learning. Through this positive process of self-reflection, athletes learn to access, interpret, and use positive feedback about what to do, not what to fix, to improve future performances.

ARCHITECTURE OF LEARNING

Highlights from the book *The Architecture of Learning: Designing Instruction for the Learning Brain*, by Kevin D. Washburn

Instruction design should be seen as different from lesson planning, a term traditionally used for pre-instruction preparation. Instruction design reaches beyond planning what content is going to be shared with the students. Designing takes into consideration a sequence of activities that are compatible with how the brain processes new information and emotions including understanding students, knowledge of the nature of learning, and awareness of subject matter.

Combining these elements requires more than traditional lesson planning. It needs instruction design that mirrors the brain's connection to learning. Robert J. Marzano stated, "It is perhaps self-evident that more effective teachers use more effective instruction strategies." A teacher can know his topic, but are they aware of how to present the information so that it is compatible with how the brain best learns?

Washburn states that the term "teacher" implies learning. If students do not learn, the teacher has been whistling in the wind." He went on, "Studies have shown that to be effective, a teacher must align instruction methods with a learning cognitive process, the brain's way of constructing understanding and forming meanings."

Washburn points out that the executives at Google say that the next big thing is "Where everything will be connected – a connection that exists in a healthy brain." The human brain's network connects everything in a behind-the-scene world of awareness and adjustments that support the nature of learning. This network seamlessly decodes, encodes, processes, and recalls information.

Washburn's Five Elements:

1. Experience–gives our brain raw data

2. Comprehension–allows the brain to sort, label, and organize sensory information

3. Elaborating–lets the brain examine information for patterns; recall prior experiences, and blend new and past experiences to construct understanding

4. Application–lets the brain use new knowledge

5. Intention–allows the brain to use new understanding in a wide range of contexts

Individuals can learn a lot more than information in student-centered/brain-compatible learning environments including coping skills, accepting responsibility, self-discovery tools, intellectual and emotional confidence, approaches for internalizing, gaining insights for challenges and pride in what one is doing to grow, all while developing self-sufficiency. Individuals grow into intellectual gym rats when guided by playful, stress-free methods. Washburn, "One of the marvelous things about the brain is that it is capable of taking over for us." The process of learning is profoundly influenced by the culture of the environment in which that information is offered.

Nature's plan for learning collaborates with real-world environments, using approaches that could be called restless, playful experimentation, which enlarge our imagination and expand our sense of the possibilities.

As a teacher, I feel my job is to make students feel capable and emotionally safe while giving them the opportunity to do some rethinking and develop new insights. A student-centered/brain-compatible learning environment is designed to change student's current insights; it is not trying to get something right. Perhaps this view is counterintuitive, but it is useful for helping students to discover and invent skills and knowledge that are personal in nature.

A student's new insight can make them aware of new information. This allows a student to be self-guided in the direction of the kind of self-learning that provides texture, which provides clarity and brings on the kind of understanding that causes enlightenment. Self-learning is clearly our most efficient student-centered/brain-compatible developmental tool.

A MICHIGAN STATE UNIVERSITY STUDY

"What makes great teachers?" Any number of factors but some are harder to pinpoint than others. A contemporary study from Michigan State University attempts to define one aspect of great teaching by focusing on how some educators use creativity to inform students.

"The importance of creativity and the need to develop critical and creative thinkers in the 21st century cannot be denied, and with that comes a requisite need for research on successful, creative teaching practices," write authors and Professors of Educational Psychology, Danah Henriksen, and Punya Mishra.

The study offers another practical purpose; "There is a strong body of thinking in educational research that essentially equates effective teaching with creative teaching." While the term "creativity" carries a certain vagueness to it, the study – using educator interviews and historical research definitions defines creative teaching as an approach emphasizing novelty (especially from the student's perspective), task appropriateness, and usefulness. As one educator told the researchers, "It is got to be effective for learning, otherwise it's just entertainment." The educators interviewed also emphasized creativity as a mindset, rather than a one-time approach to a lesson.

The M.S.U. research uses sampling from 90-minute interviews with eight recent National Teacher of the Year finalists (only first names were mentioned in the study). The following are among the responses that the teachers gave in their discussions:

Creativity is learned, not innate: "You have to be able to move from the concrete to the abstract and back again, to synthesize."

Teachers can draw on their own interests to introduce creativity into the classroom: "What I do is basically, I just go through life and always – I am always on the lookout for how I can apply that to teaching."

It is not the teacher that necessarily needs to be creative. Instead, teachers can 'let loose the reins' and allow students to be thoughtful; "As a 32-year old, my brain has been taught to think in a specific way with different things. Their 8-10-year old brains are a lot more open."

Take risks, but plan through them; risk taking should not be haphazard; forgive yourself when things do not go as you might have hoped.

Lessons should carry real-world value. One science teacher, for example, described creating a town-hall environment in his classroom where some students represented different interests in the energy community, and other students acted as politicians trying to put together an energy plan.

Use cross-curricular approaches. One math teacher draws on business and psychology to help students understand the math behind sales pricing. "The students get very engaged in advertising and how advertisers try to target them as young adults."

The researchers offered a caveat about real-world learning, in that it does not necessarily imply creativity, but that a creative teacher might be more willing to take risks that draw students into real-world settings.

Another thing that the researchers brought up is that federal and state policies may inhibit the ability of teachers to feel like taking creative risks, especially in schools that utilize "teacher proof" curriculums in preparation for standardized testing.

One of the teachers interviewed celebrated administrators who gave educators the freedom to try new things. The study suggested that professional development can help teachers develop creative approaches. Schools of education can also emphasize creativity to pre-service teachers.

"When teachers are deprived of the opportunity to foster creativity in their classrooms, students cannot begin to develop a mastery of critical or creative thinking abilities" the study concludes. Perhaps even worse, because of downstream effects, teachers through sheer repetition of giving the same lessons, become bored and lose enthusiasm in what they teach, an attitude that is certainly visible to the students and affects their own enthusiasm and interest in the subject matter.

THE PURSUIT OF WHAT WE DON'T KNOW

Dr. Stuart Firestein, Columbia University, neuroscientist, teaches a class called IGNORANCE 101. He points out that science is best seen as uncovering "high-quality ignorance," which is what else can be learned. Dr. Fierstein, in his TED talk, mentioned that scientists talk about what they do not know when they get together. The pursuit of what we do not know, the aim of learning, is being left out of teaching and fixing - lessons that are just providing information.

In a *New York Times* review of the book *The Island of Knowledge*, by Marcelo Gleiser, "It is a misconception that scientists are interested, above all, in getting confirmation of their theories; not so. They are pleased if their point of view has been found to be incomplete. This gives scientists the opportunity to learn new things. Unfortunately, it is not uncommon when science uncovers that a long-held theory is incomplete, that such findings have not stopped people from repeating the same argument."

Dr. Stephen Yazulla would tell the members of his lab that since they did not have a role in designing the brain, they should have no vested interest in how their experiments turn out. It was most important that their results were beyond reproach; then we can all argue about what they mean. As long as all agree on the data, it is information first, then interpretation. When it comes to designing a path to learning, educators and sports instructors should keep this insight about scientists in mind; it opens opportunities.

The following was sent to me by Gareth McShea, one of Ireland's leading PGA instructors.

> "Perfectionism is not self-improvement. At its core, perfectionism is about trying to earn approval. Most perfectionists grew up being praised for achievement and performance (grades, manners, rule following, people pleasing, appearance, sports). Somewhere along the way, they adopted a dangerous and debilitating belief system of "I am what I accomplish and how well I accomplish it. Please. Perform. Perfect."

From *Daring Greatly*, by Brené Brown

> Healthy striving is self-focused: How can I improve?
> Perfectionism is other-focused: What will they think?
> Perfectionism is a hustle.

INTEREST BEFORE FUN

Keeping students interested by giving them choices and allowing them to use their own curiosity and imagination, births meaningful learning. Results from research show that it is interest that brings on the fun of learning. When students are not making progress, but what they are learning still holds their interest, fun and progress can follow.

The human brain is a sense-making organ that is drawn in the direction of what interests its natural curiosity. The fun involved in learning precedes the student's interest in solving the problem. Sports psychologist Dr. Rick Jensen has pointed to the value of training techniques that keep students interested and curious.

Some learning environments are not efficient engines of change because personal insights get lost in someone else's corrections. The joy of learning anything can be experienced by linking up what the science of learning is uncovering about approaches to education.

Effective managers and educators build confidence in the power of curiosity, imagination, and what-if ideas of others. They help individuals find personal insights for problems to be solved. Allowing the illusion of how-to directions from an expert to take over a learning environment only slows down gaining knowledge that is personal in nature for future use.

We all come into this world curious to learn with an urge to invent survival skills and develop our own information base. To have the mind energized by an interest and the urge to keep looking may be more valuable than a positive outcome. The energy of anticipating what's next can be stronger than the satisfaction of completing a task.

CURIOSITY

Where does curiosity come from is a topic of *The Hungry Mind: The Origins of Curiosity in Childhood*, a book by Susan Engel, Professor, Williams College

Most important scientific breakthroughs have come about because someone felt a sense of intense curiosity - an irrepressible urge to find something out. Life is filled with mysteries begging to be solved by each of us.

Alexander Fleming discovered penicillin when he noticed that some of the cultures in his lab had succumbed to a fungus but others had not. He just had to know why. That first spark of confusion and curiosity - when things are not as expected, lies at the heart of world-changing findings like Fleming's. The same applies to journalists, determined to uncover the facts of a story (for example, Bob Woodward and Carl Bernstein, of Watergate fame), or Victorian novelists such as Mary Ann Evans (aka George Eliot) who felt compelled to find out what went on in the neighbor's bedroom. The urge to know doesn't first occur in the college lecture or even the lab; it begins in the crib.

Once babies become children, their curiosity grows much more sensitive to the environment.

A frowning adult; questions asked but unanswered; or a classroom devoid of complexity all have the power to discourage a child's exploration which helps explain why curiosity seems to dwindle as we age, changing from a weed to a fragile plant. Every toddler is curious, but not all adults are.

The research in my lab shows that far from nurturing curiosity, schools seem to repress it. The pressures to deliver information, hone skills, stick to the plan and avoid the unknown work against a child's natural curiosity. However, it need not be so. Classrooms could be green-houses for curiosity. Questions could be encouraged and guided, exploration could be at the center of the curriculum, and rather than being pushed to the side, children's specific interests could be fostered. Given how central curiosity is to learning and human progress, why not cultivate it? After all, it is our other mother.

THOUGHTS

Mental models drive the decisions that we all make. Everything begins with a thought, conscious or non-conscious. The implication is that to support successfully gaining a meaningful education, there are times when the mental model that some educators and students use needs to change. Said another way, improving mindsets first supports improving skills sets.

Steve Jobs once said, "People do not know what they want until you show it to them." What follows demonstrates why we should want to use a brain-compatible/student-centered approach for learning; it is a methodology rather than just a set of outcome goals.

A student-centered/brain-compatible process draws learning out of students, filtered by their experiences.

A student-centered/brain-compatible approach is a process that acts less as a teacher who is just looking at and criticizing outcomes and more as a mentor who keeps a student's mind active; a lens through which meaningful learning is gained.

Student-centered/brain-compatible approaches realize that students first translate what is being shared through their emotions.

When students are first given a big concept or general picture of the subject, its parts will later fall into place. Starting with the details or minutia leaves the student in the dark as to what the context is. Details, in the absence of the larger picture, hang like a chandelier in the middle of the room without any other support. Think of an hourglass, start with a broad topic, narrow it down to a detail and then expand again to put the topic back in the original context.

"Entrenched ways of doing things can be tough to change. Without going into the pros and cons of the issue, an example is a huge controversy over the implementation of the Common Core in grade schools." (Dr. Stephen Yazulla)

Traditional best practices from experts often do not accommodate changing environments.

> **A brain-compatible/student-centered learning process addresses unanticipated problems with flexible knowledge and portable skills.**

RECAP

Being uninformed about the nature of learning has been found to be a wide-spread problem. When it comes to the topic of learning, there are no mysteries, only things to become aware of. You must be aware of the elements found in the nature of learning in order to put them to use.

The learner's brain wires itself in reaction to the learning environment it is experiencing.

Human beings have the use of prior knowledge and the internal operations of the brain that provide support for experiencing meaningful learning.

With "brain in mind" information delivery, processing information, situation adaptability, abstract thinking, creativity, visual recognition, organizing skills, fluid intelligence, flexible thinking, skill storage, emotional stability, continuous learning, performance management, talent mobility, reasoning and deduction skills, imagination and long-term retention are all available.

The components of efficient positive assessment include learning orientation, authentic, validity and social assessments. Positive assessments support students to apply the necessary what-to-do (not what-to-fix) knowledge during the student's own learning process.

YOUR NOTES ____

____

____

CHAPTER NINE

Emotions

The acquisition of any skill is critically dependent on the physiological, hormonal, and neural conditions of the brain at the time of learning. A "Brain-compatible/student-centered Learning System" takes these factors into account, trying to provide conditions in which learning is stimulated rather than inhibited." (Dr. Stephen Yazulla)

"Emotions are the rudder for thinking." (Antonio Damasio, Brain and Creativity Institute, University Southern California Neuroscientist)

"BIRDS FLY, FISH SWIM, PEOPLE FEEL." Mary Helen Immordino-Yang, Ed.D.

EMOTIONAL CONNECTIONS

It's been said that the basic purpose of emotions is to keep us alive. If we see something as beneficial, we approach and accept it. If seen as a threat, we move away or move the threat away. The relationship between our emotions, bodies and thoughts are fundamental to how and why the brain reacts and operates the way it does. These emotional connections point to the reason that feelings, thinking and the body are inseparable. Our emotions can positively support or negatively suppress our ability to reach our potential. The important role the emotions play in learning is based mostly on the differences between a teaching-fixing to get it right path to learning and a learning-developing-growth approach.

> **Any idea exchanged between two people is an emotional experience, a social interaction that can support or suppress learning. Unfortunately, in teaching-fixing environments to get it right, the students often hear "you will never make it unless you do this or that" or something similar. How do you think students feel after hearing that?**

Emotions are a huge component of our ability to experience meaningful learning. How students feel about the components of where, how, from whom and what they are learning matters. When educators lead by making a student's emotional response their first concern, meaningful learning can become a reality.

Learning is an experience. All experiences are emotional. "The way we feel influences how we learn in every learning environment. We attach feelings to things and things to feelings. We always feel something that pushes us on, or away from what we are learning."(Mary Helen Immordino-Yang, Ed.D., a leading researcher in the field of the emotional influence on learning)

We have emotions that bring attention to, or away from what we are learning. The emotions of fear and stress are mostly involuntary, unconscious reactions based on one's experiences. Fear and stress HURT learning and performing. Meaningful-learning environments are emotionally safe and devoid of negative judgments of outcomes.

Because emotions are a critical component of meaningful learning, how individuals see themselves when learning often has to change. Carol S. Dweck, Ph.D. refers to how we see ourselves as a *Mindset*, the title of her must-read book.

For example, when learning, some individuals see themselves as, "I have something missing," or "I have things that need to be fixed." On the other hand, individuals who see themselves as who they are at that moment are in place to start a journey of new learning and developing. They do not see themselves as broken, in need of fixing or having something missing. That no one is broken, in need of fixing or missing something is an important mindset for both the instructors and students in any learning environment.

The notion we should first make, and the knowledge we currently have, should be seen as who we are. Then, in the future, we can learn to use first notions, or our current information base, in different ways. When learning, our perceptions, balance, rhythm, and timing should be seen as developing and not in need of being fixed.

The nature of learning is the nature of human development, a process that is fueled by an engine that is doing more than just gathering subject content information. It is an engine that is fueled mostly by how individuals see themselves when they are learning, which is heavily influenced by the design and structure of the approach to learning that is being used by the instructors and students.

Cognitivists have proposed various models of how emotions can structure, guide, and influence mental representations. These models point to a simple truth:

Dr. Bowker," if one wants something to be attended to, mastered, and subsequently used, one must be sure to wrap it in a context and words that engage the positive emotions. Conversely, experiences devoid of positive emotional impact are likely to be weakly engaging and soon forgotten leaving nary a mental representation behind."

> **"Creating an educational environment in which pleasure, stimulation, and challenge flourish is an important mission. Also, students are more likely to learn, remember, and make subsequent use of those experiences with respect to which they had strong positive emotional reactions." (Dr. Matthew Bowker)**

"Learning is deeply affected by the emotional environment in which learning takes place." It has become evident that any portrait of human nature that ignores motivation and emotion will be of limited use in facilitating human learning." (Dr. John Medina)

I find it interesting that many books about the brain's connection to learning highlight the interplay between emotions and our cognitive capabilities. The reality is that emotions can help or hinder acts of learning. In the center of our brain lies the limbic system; this is called the emotional brain. Richard J. Davidson, Ph.D. pointed out that the pre-frontal cortex of the brain is also involved in emotions in his must-read book, *The Emotional Life of Your Brain*, written with Sharon Begley.

Emotions are chemicals. When we say we like something, the chemicals that are released reinforce that emotion. Every cell in the brain responds to chemicals that can produce an emotional component. Everything, including learning, starts in our cells and the chemical exchanges in which they are involved.

The brain is the gateway to learning on a path of emotional experiences. When we are learning, the body, mind, and brain come together emotionally. It may help to see the brain as having information and the mind as filled with emotions that are learned after birth, influencing our ability to learn, develop, and grow as we move through life. Telling someone he has failed is perhaps the MOST discouraging and potentially dangerous emotional message there is. It should be recognized that emotions can never be fully suppressed and that is why providing meaningful feedback should not be about what was wrong, but about what could be different.

> **All learning is filtered through prior knowledge and the emotional state of being and motivation to learn." (Andrea McLoughlin, Professor Long Island University)**

The emotions of "I need," "I just want to," "I should," can derail progress. There are layers of performing based on the story we tell ourselves. Our perceptions and feelings are formed by what we are looking for, expect to see, or what we want, all based on our experiences.

"We think in the service of emotional goals. We think about what matters to us."(Immordino-Yang and Damasio, *We Feel, Therefore We Learn*. J. Compilation 1: 3-10)

Emotions are involved in organizing brain wiring; what seems good or bad, wanted or avoided. This is created by the concepts and feelings generated by emotions. The research of Mary Helen Immordino-Yang into emotions and learning has been groundbreaking. She points out that when learning is an emotional event for students, instructors should keep this in mind when they are delivering information and evaluating student's performance. How does the student react emotionally to what is taking place? Do they feel safe or threatened? When helping individuals learn with an approach that does not include negative judgments of outcomes, the emotions of frustration or intimidation are lowered or vanish.

> ***Suggestion!*** **exchange traditional and at times emotionally polarizing teaching-fixing to get it right approaches to learning for galvanizing learning-developing approaches that are brain-compatible with emotionally safe learning. There are 50 shades of the color white, and there are just as many, if not more, stress-free ways to connect to learning.**
>
> Emotions can be positively or negatively influenced through:
>
> Self-regulation
>
> Seeing options
>
> Problem-solving skills
>
> Memory
>
> Heart rate
>
> Learning potential

Brain-compatible/student-centered information-delivery systems do not create the kind of stress that negatively influences emotions. Learning is mostly an emotional event, and brain-compatible/student-centered learning approaches take this reality into consideration.

It has been found that cognitive performance suffers in environments that have the emotional components of excessive stress, fear, or social judgments. These environments compromise the neural process of emotional regulation, causing the brain to go into a "protective" mode, closing down some of our neural pathways.

Research from cognitive science and other disciplines shows that the nature of learning is influenced by the brain's limbic or emotional system as information travels through the self found in self-discovery, self-organization, self-assessment, and self-confidence, to name a few of mankind's many self-skills. New learning is or is not encoded while physical and emotional safeties are being evaluated.

According to Harvard's Connecting Mind-Brain to Education Institute, it is very important to realize that one of the brain's natural tools is self-protection, which can reduce the number of neural pathways available for learning when we are under excessive emotional stress. This causes learning to become less efficient with fewer neural pathways available. The negative side effects of fewer neural pathways available throughout the brain include depression, confusion, reduced concentration, less efficient vision, poor spatial thinking, and poor working.

TOPIC: THE BODY, MIND, AND THE SELF

The learning process is social, psychological, & emotional. We feel, therefore we learn. Birds fly, fish swim, humans feel. What follows are notes I made during a lecture that Mary Helen Immordino-Yang, Ed.D. presented at Harvard University's Connecting the Mind-Brain to Education Institute.

> Learning is influenced by culture and customs. All background information influences new learning.
>
> Having compassion for pain, admiration for skills and admiration for virtue supports learning.
>
> Emotions change the whole body-brain system by releasing chemicals.
>
> Changing the way people come to learning, it starts in the brain.
>
> Learning is not about information. It is about developing the ability to see options, using and constructing knowledge personal in nature.

Coverage - Just giving out information is referred to as coverage and this does not develop meaningful new learning.

Motivation - Intrinsic motivation or self-motivation and learning for the fun of it is more useful than extrinsic motivation, or learning for what you can get, or learning to please others.

Have students explain what they are learning and how it can be used in several different ways; this will help to encode long-term meaningful learning.

Make learning environments emotionally safe.

In John Medina's book, *Brain Rules*, he noted the following insights from his research:

Emotions make the brain pay attention. (p. 256)

Our ability to learn is based on a series of increasingly self-connected ideas. (p. 270)

We are always using pre-loaded information. (p. 267)

New information "re-sculpts" prior information and sends the re-created whole back for new storage. The new information and the past information now act as if they were encountered together." (p. 129)

Emotions are a huge component of our ability to experience meaningful learning. How students feel about the components of where, how, from whom, and what they are learning matters. When, during learning opportunities, educators lead by making a student's emotional response their first concern, meaningful learning can become a reality in S.A.F.E., PLAYFUL environments.

A SAFE EMOTIONAL ENVIRONMENT

Many questions about learning are grounded in asking if the strategy for learning used by instructors and students promotes motivation and interest, or does it cause frustration and intimidation? Instructors and students should keep this question in mind when they approach learning. Everyone has the right to feel safe in learning environments. Individuals cannot do what they are capable of, or be who they are if they are confused, frustrated or intimidated. It is a nurturing force, not fixing and criticism, that supports lifelong learning. Telling students what is wrong or what needs fixing is not as brain-compatible as guiding them in the direction of learning what to do, without telling them how to do it.

When frustration and intimidation are lowered, or removed, from approaches to learning, clever, creative, competent puzzle solvers with the tools to adjust and cooperate with ever-changing environments emerge.

Our fight-or-flight competitive instincts are not as important to our evolution and development, as the skills of cooperation. This insight was pointed out by Michael Tomasello, in his book *A Natural History of Human Thinking*. Teaching-fixing is an authoritarian, not a democratic method of learning, free of intimidation that fosters competition, not co-operation when learning.

> **By removing frustration and intimidation from learning environments, approaches to learning become more brain-compatible. It is accepted that there is no silver bullet or single concept that will enhance approaches to learning. When students are made to feel capable; this comes close to being a silver bullet.**

When the emotional conditions that favor meaningful learning are understood, individuals learn faster and with more depth of reasoning than they had previously believed possible. As I have said before, for an instructor to successfully stimulate learning and growth in others, it helps to have insights into the nature of learning. If you have been brought up in a culture that believes that learning is difficult, and poor outcomes need to be fixed, you may want to consider using a brain-compatible/student-centered approach to learning.

Emotion plays an important role in all facets of learning. "I like," "I do not like," "I can't do this," "I can do this," "this is hard," " this will be easy," are all statements based on the emotional story people tell themselves in learning environments. An important insight: emotions pop-up at every stage of learning, improving and performing.

The stories we tell ourselves about our interaction with the environment will, either positively or negatively emotionally influence the outcome of the learning experience.

Self-growth, self-confidence, and learning are supported by an internal conversation from a curious inner emotional voice within our brain. On the other hand, the inner voice of self-doubt can damage a student's self-confidence and self-growth. I have found that keeping curiosity alive is a most effective motivational tool for reaching one's potential. People are motivated when they are curious.

Our inner voice, the story we tell ourselves when learning and performing, is either brain-compatible or it is not. The tone of our self-talk, positive or negative, is influenced by the design and structure of the information delivery system that is utilized. Again, how information is delivered to students by their self-talk or from other sources MATTERS.

THE POWER OF WORDS

"IT IS NOT WHAT WE SAY, IT IS HOW WE SAY IT" is a well-known insight; one that pertains to experiencing meaningful learning. The view here, as shared by Professor Terry Doyle, is that with a student-centered, brain-based approach to learning, teachers should not give commands; they should cause students to think and create.

How words are used may top the list of elements that create emotionally safe learning environments. Learners subjected to criticism, or suggestions that they were unsuccessful, do not respond as well as those who receive positive encouragement to experiment further. "Words are the most powerful performance-enhancing drug known to mankind." (Rudyard Kipling)

The brain links words and emotions with our cognitive skills, and this has an important influence on acts of learning. Thoughts, words, and conversation are at the heart of everything we do, including learning. When delivering information to ourselves or others we use words, hopefully, they are compatible with how the brain best learns.

WORDS CAN cause every kind of emotion:

be remembered or forgotten	be creative or destructive
be creative or suppress interest	be true or false
be warm or cold	hurt or support
be proactive or reactive	make sense or create confusion
be to the point or be evasive	describe impressions and insights
be on time or too late	be ours or someone else's

We all have heard, "it's not what you say, it is how it is said that's important." By using words and strategies that support the nature of learning, instructors become more effective. These instructors are making students more aware of how their ability to learn works. In brain-compatible/student-centered learning environments students are often "students in training for learning," guided in the direction of developing and inventing their own best approach to their potential.

When helping someone to learn and improve, use "guiding words," not criticisms or judgments. Guidance uncovers possible solutions by providing information that inspires new insights. Guidance also avoids trying to fix poor habits and reacting to unworkable outcomes.

What we say, how we say it and when we say it can enhance learning and performance. Words can also be destructive. Words and the way they are being shared can inflict damaging emotions, including self-doubt. By examining how we respond and communicate, we can become more skilled with the use of words when learning, teaching, and performing.

Now think about the learning environment and ask yourself what kind of words from guides and coaches have consequences that can support or suppress learning? Was there a difference?

While keeping the use of words in mind, note what Matthew H. Bowker, Ph.D., a researcher who explores critical thinking stated, "The relationship between the instructor and student must be caring, equitable and responsive. It must be firm and safe. The tone must be playful and creative so students can think, converse, listen and question without feeling either lost or crushed.

STUDENT'S WORDS

Meaningful approaches to learning speak to the internal, emotional self of individuals. How we talk to ourselves and the words others use when we are learning have a profound influence on being "good accomplishers," a Joe Dispatie term. Our personal perceptions, based on our experiences, are responsible for how the brain is wired at the start of our interactions within changing environments.

When a student speaks, of course, the educator or coach will sit quietly and listen until they finish. That's not where listening ends. That's where the listening process begins, and the coach then has an active role to play.

After hearing and processing the student's words, the instructor formulates an interpretation of what the speaker said. It is then critical that the instructor clarifies that what s/he heard was what the student intended, by summarizing to be sure the communication is correct.

In response, the student either acknowledges the accuracy of the interpretation or rephrases/ re-states the information, and the process begins again until the student and teacher achieve a common understanding. Once there, the co-creative learning relationship is fully engaged.

Whenever an educator or coach fails to clarify their interpretation of the student's words, the process breaks down. The resulting conversation will be based on assumptions, not reality, and the student has not been fully heard and understood.

Active listening is a skill which like any other skill, improves with practice. So, what is the secret to being an effective educator? Practice actively listening until it becomes a habit.

Each of us has a personal filter through which verbal information flows. Our filters are based on our unique life experiences, education, culture, religion/spirituality, language, work/career, etc.

A speaker's words are delivered through his/her personal filter. The listener hears

those words through his/her own filter (created from their own life experiences, education, culture, etc.). In a coaching session, the Client is the speaker, and the Coach is the listener.

NERVOUS SYSTEM FIRST

Everything we have seen, heard, or physically and emotionally felt and thought about, was experienced somewhere in our nervous system first. The central nervous system (CNS) receives information from our senses, intellect and memory and then forms reference points to be used during future acts of learning and other interactions in life. Everything starts at the cellular level of our being. The brain organizes the program first, then our actions and other thoughts follow.

When individuals are inexperienced in a topic area, they can become overwhelmed by new information. Novices to a topic area have few experiences to draw from and are therefore are less able to or sometimes unable to give meaning to and make sense of new information; two requirements of new learning.

There are studies from UCLA's Learning and Forgetting Lab and other research centers that show the outcomes arrived at during lessons, both workable and unworkable, are not indications that meaningful long-term learning has or has not taken hold. A counterintuitive insight, but true. Using the same information in different environments and receiving a workable outcome would show that learning has been encoded.

In 2001, Benjamin Bloom's steps of learning were revised by Anderson and Krathowohl as follows: (note they are brain-based steps)

Recall goes to Understanding

Which goes to Application

Which goes to Analysis

Which goes to Evaluate outcome

Which goes to Creating the next outcome.

BEHAVIOR

What comes first, our behavior, or our intelligence? It's our behavior, followed by advancing our intelligence. In the book *On Intelligence*, Jeff Hawkins and Sandra Blakeslee point out that the brain operates by prediction: if this, then that. Fortunately, when it comes to learning, this internal brain activity is now understood more fully than in the past.

"The single most important factor in learning exists in the learner's neural network (what they know); we should determine what that is and then proceed accordingly" Cell biologist James Zull, author of *The Art of Changing the Brain*.

For me, information becomes knowledge after it can be put to use for delivering workable outcomes. The use of metaphors, stories, and analogies in learning environments supports this process. For example, John Pollack's latest book *Shortcut: How Analogies Reveal Connections, Spark Innovation, and Sell Our Greatest Ideas* highlights several important cognitive breakthroughs due to analogies.

Pollack points out that Thomas Edison, the Wright Brothers, Charles Darwin, a Henry Ford employee Bill Klann and Steve Jobs all made history by learning through analogies. To make an analogy we use tools of learning, making a comparison and parallels between two distinct things, explicitly or implicitly. Learning to be nimble at finding parallels is supported by brain-compatible/student-centered learning environments.

For example, the elements of balance, timing, and rhythm of tossing an object underhand, exist in everything we ask our bodies to do.

When an approach to learning is compatible with the Nature of Learning some of the qualities include:

Emotionally supportive (many are not)

Free of negative criticism (few are)

Avoid judgments (most do not)

Developmentally appropriate (many are not)

Introduce information or skills that are just beyond current skill levels (few do)

Find unworkable outcomes useful for improving (most do not)

Give students choices (few do)

Change poor insights (few do)

Physically safe (some are not)

The research of Dijksterhuis and his colleagues found that "unconscious thinkers" outperformed conscious analysis when making decisions. It was shown that unconscious thought produces better decisions than when people decide by using conscious, logical reasoning. Taking a break from a task that demands attention, improves problem-solving. (cited in Ringleb and Rock, 2013; 138) (Athletes take note!)

"People are generally better persuaded by the reasons which they have themselves discovered than by those which have come into the mind from others." (Pascal, Pensees, p 86) New learning is based on one's experience, and that connection forms a hidden and often non-conscious language of insider-information that supports the experience of long-term learning.

THE ROMANCE STAGE

Benjamin Bloom called the stages of learning the early, middle, and late. These stages parallel what Alfred North Whitehead in 1929 referred to as the **"Romance Stage,"** the **"Precision Stage,"** and the **"General Stage."**

Studies about human learning and development point out that what happens during the early or romance stages of learning influences how long and how often individuals will stay involved with an activity.

The early romance stage must be recognized for what it is. This stage is when the possibilities of continued engagement and the possibility of improved skills exist. This is the stage when passion and a love of something can be nurtured. If we want people to fall in love with an activity, then this awakening or arousing stage must be free of intimidation and frustration, (this is one of the most useful insights in the book).

Studies show that without interest and passion there will be no love for playing a game or learning a subject. Without interest and love, the basic needs of human behavior and development are not being met. "The brain only learns what it pays attention to."(Professor Terry Doyle)

Skill development is a byproduct of free, self-motivated play. Pick-up games in parks and schoolyards, sports environments, and classrooms free of judgment and criticisms are all self-motivated play. This environment should exist in any other learning environment.

The freedom and enjoyment that fosters interest and passion for an activity are being suppressed when the early (romance) stage of learning is filled with required structured training, lots of instruction, and competition.

Over-managed, and structured instruction during the early romance stage of learning can cause individuals, young and old, to become frustrated and intimidated, thereby missing an opportunity to develop interest and passion for a subject and love of a game, or a love of a subject, or the process of learning anything.

I have learned that it is important to recognize that without the sense of pure enjoyment during the early stages of learning, there will be a lack of persistent long-term interest. Without enjoyment and self-confidence during free play, there will be little opportunity to gain a feeling of self-worth and love of a game or subject during a journey of development.

The nature of learning will find a place for a little structured training during the early or romance stage, as long as there is the opportunity to experience much more free play filled with exploration.

> ***Suggestion!*** **When learning and playing for its own sake, students may develop a passion for and fall in love with that game or school subject regardless of the score or subject to be learned.**

Studies show that people engage in playful experiences all the time, without anyone telling them to do so. All humans choose to be playful; it is our nature! When it comes to sports and subjects, people of all ages stop playing or play less frequently because of boring practice sessions, not enough free play or engagement, emotional stress from excessive performance demands, negative coaching, and a feeling of failure. Coaches might note that drills, when first learning, are of less value than after one knows how to do something.

It seems imperative that before individuals fall in love with and become frequent participants in learning anything, they must experience an introduction stage (aka, the Romance or early stage).

> **Learning becomes more difficult than it should be when information is delivered as a "subject-matter experience" and not as a playful "talent-management, developmental process."**

CREATIVITY

Brain-compatible/student-centered approaches to learning support creativity. Sir Ken Roberson during his TED® presentation said, "Creativity is not only about generating ideas; it involves making judgments about them. Testing and refining them and even rejecting some."

When it comes to the nature of learning and creativity, knowledge is not just for consumption; it is meant to be leveraged into a formative experience. Learning and creativity in brain-friendly environments are like designing a car while you're driving it, with a fluid mixing of ideas.

Creativity moves learning in the direction of basic brain-compatible/student-centered learning, generating knowledge ideas. Creativity enables the learner to move away from just consuming information.

In her book, *Sparking Student Creativity*, Patti Drapeau points out that Westerners tend to think of creativity as a novelty with an emphasis on unconventionality, inquisitiveness, imagination, and freedom. While Easterners think of creativity as connections between old and new knowledge. When the approach to learning is brain-compatible, both these views are included.

Creative thinking is referred to as flexibility, fluency, originality, and elaboration. I suggest this kind of thinking is supported by brain-friendly approaches to learning. This insight recalls the kind of "flexible knowledge and portable skills" that Professor Daniel Willingham points out support the nature of learning and its connection to the brain. Flexible knowledge and portable skills have been at the core of the nature of learning and our survival skills from day one.

"We have managed to survive nearly every climate change our planet has offered (so far). Ten thousand years ago, humans, pets, and livestock accounted for about 0.1% of the terrestrial vertebrate biomass on earth; we now account for 98%. Our success is in large part owed to our cognitive capacity; the ability of our brains to flexibly handle information." (J. Levitin, *The Organized Mind*; 8) The Creativity Site is a website worth visiting.

EFFICIENT COACHING

I found the following about coaching on GetSportsIQ.com. I believe that where it states "athlete," the terms student, employee, or child are interchangeable.

Managing approximately 70 gymnastics professionals, all of whom are teacher-coaches, I am acutely aware of the amount of training and education that these dedicated pros undergo to instruct their young athletes. The technical knowledge of the skills in combination with understanding the progressions necessary to achieve the elements safely and the rules and regulations that govern the various competitive levels fills volumes of books, hundreds of DVDs, and dozens of training seminars and conferences.

But that is only part of the picture. While superior knowledge of the sport is a cornerstone of an efficient coach, it takes so much more than content and procedural knowledge to be an efficient coach or teacher. Simply because a person has great knowledge of the sport and a fabulous win-loss record, does not mean they are an efficient coach.

Efficient coaches...

Cherish the person over the athlete. Efficient coaches know that being an athlete is just a small part of being a person. Efficient coaches never do anything to advance the athlete at the risk of the person.

Listen to their athlete's concerns. Efficient coaches don't tune out athlete's worries, fears or mentions of injury.

Connect before you direct. Efficient coaches understand the importance of emotional connection. You matter. You belong. You are important to me. Not you the athlete; rather, you the person. Our most fundamental need is emotional and physical safety. When we feel safe, we can trust and when we trust we can learn. Efficient coaches know that this foundation of trust is essential.

Begin with the end in mind. Efficient coaches keep their attention on the big picture of the goal of the athlete. They have a plan, but are flexible, as they are aware that the road to success is filled with twists and turns.

Embrace athletes' struggle. Efficient coaches understand that learning is a curve. Like muscle needs to break down before building up, athletes need struggles to push forward. An efficient coach doesn't panic when this struggle happens.

State corrections in the positive. Efficient coaches say "do this," rather than "don't do this." Don't bend your arms is less effective feedback than "push your arms straight."

Find the bright spots and build from there. Efficient coaches are aware of what needs development, and try to improve them to meet minimal standards but spend much more attention on the areas that an athlete excels in. Trying to turn a strong pitcher into a better batter is less effective than trying to make him better at his curve ball.

Don't try to break bad habits; rather, build new habits. Efficient coaches know that the most effective way to break a bad feedback loop is to replace one habit for another.

Be careful about how you measure success. Efficient coaches do not use scores or win-loss records as their sole measure of success. Efficient coaches understand that doing so can erode the long-term development of the athlete. Efficient coaches instead develop competencies for the long run, even if that means sacrificing success at the beginning of the journey. If you had to choose, would you rather have your child be the strongest student in the first grade or in the twelfth grade?

Use the right mixture of attainable and desirable goals. Efficient coaches have zoned in on the sweet spot of challenge.

Constantly seek continuing education. Efficient coaches never believe they know it all, or that they cannot improve themselves. Quite the opposite. Efficient coaches read journals, articles, books and scour the internet for training ideas. They attend professional workshops and seek mentorships from other coaches.

Create, instead of finding talent. Efficient coaches appreciate natural aptitude but know that it can only take an athlete so far. Furthermore, efficient coaches are humble enough to admit that they are not perfect at predicting success, so they just get in there and work.

Efficient coaches concede that extraordinary talent is not a fair assessment of their value as a coach; rather, they measure their coaching efficacy by taking an athlete who is less gifted and helping that athlete succeed. How many beginners that we coached are still playing years later, can answer how effective a coach we are.

Observe intently. Efficient coaches are always trying to figure out what makes people tick so they can better reach them.

Understanding interpersonal relationships of the team is important. Team building and bonding are not wastes of time but essential elements for success.

Separate training and learning from practice. Efficient coaches understand that practice begins after the athletes learn. Thus, they do not have athletes practicing something they have not yet learned, to avoid creating bad habits. Training is acquiring; practice is applying.

Focus the athlete on what to do, not what to avoid. Efficient coaches tell their athletes things like "Shoulders squared and body tight.," versus saying, "Don't fall."

Understand human development. Efficient coaches have a working knowledge of the milestones of human development and tailor their actions and expectations to meet the athletes where they are.

Use positive coaching techniques. Efficient coaches do not yell, belittle, threaten, or intimidate. They do not need to bully to get results. While short-term success may occur under such pressure-filled environments, an efficient coach knows that in the long run these techniques will backfire and are dangerous to the development of the child.

Have a learning, developing, growth mindset. Efficient coaches believe that our basic skills can be developed through dedication and work. They reinforce this with their athletes over and over so that their athletes feel motivated and are productive.

Know what you don't know. Efficient coaches are not afraid to admit that they don't have all the answers. They do not allow their ego to prevent them from getting additional help, training or even suggesting to an athlete's family that the athlete needs to move to a more experienced coach.

Educate athletes. Efficient coaches go beyond instructing their athletes; instead, they educate them in age-appropriate ways regarding the purpose of their objective.

Understand that interest and fun are essential elements in training, no matter how elite an athlete becomes. The number one reason that athletes quit sports, even sports that they love and in which they are succeeding is because they lose interest and it is no longer fun. Interest and fun are not frivolous but at the foundation of an athlete's healthy commitment to a sport. Note, fun follows an interest, keep individuals interested, and they will have fun.

Content knowledge is just the beginning of what makes an efficient coach (or teacher). Yet, absent these other qualities, all the knowledge in the world does not make a smart or effective coach efficient.

"Learning is more efficient when using a plan for learning, than a course outline, and when guided by more of a process structure than content structure." (Malcolm Knowles, *Self-Directed Learning, a Guide for Learners and Teachers*)

RECAP

Emotions are a huge component of our ability to experience meaningful learning. How students feel about the components of where, how, from whom and what they are learning matters. When educators lead by making a student's emotional response their first concern, meaningful learning can become a reality.

Unfortunately, because of the strong influence emotions have on learning, people of all ages tend to hear the bad news first and loudest. Educators, guides, and coaches should not assume that they can balance or offset negative messages with positive messages.

The emotions of "I need," "I just want to," "I should," can derail progress. There are layers of performing based on the story we tell ourselves. Our perceptions and feelings are formed by what we are looking for, expect to see, or what we want, all based on our own experiences.

"We think in the service of emotional goals. We think about what matters to us."(Immordino-Yang and Damasio, *We Feel, Therefore We Learn*. J. Compilation 1: 3-10)

Research from cognitive science and other disciplines shows that the nature of learning is influenced by the brain's limbic or emotional system as information travels through the self-found in self-discovery, self-organization, self-assessment, and self-confidence, to name a few of mankind's many self-skills. New learning is or is not encoded while physical and emotional safeties are being evaluated.

Emotions are a huge component of our ability to experience meaningful learning.

Negative emotions can block the learning experience.

Negative emotions are difficult to offset.

Our perceptions and feelings are formed by what we are looking for, expect to see, or what we want, all based on our past experiencing.

YOUR NOTES ____
____
____

CHAPTER TEN

Basic Look at Memory

When learning, or performing, it is more useful to remember than to think. "Our memory is designed to save you from having to think. Our memory stores strategies to guide what we should do." (D.T Willingham, Why Students do not like School; 3)

Most of what we do and learn every day is influenced more by our memories of prior knowledge than by conscious thinking. Willingham points out that we are not that good at it, but we do like to think. We are naturally curious and look for opportunities to think. Often, we think when we should have let our memory guide us.

Information changes the brain, memories form, then the brain changes outcomes!

MEMORY

In *Scientific American Mind*, Nov/Dec 2014, the article "Cyborg Confidential," by Sandra Upson, stated that University of Pittsburgh neuroscientists recorded brain activity with a computer, and it revealed interesting insights into HOW WE LEARN. These scientists had an unprecedented view of how the brain proceeds from thoughts to actions, and how we develop new skills. The short answer to the how learning is caused to take place is by cellular mechanisms within the brain. These occur in predictable regions of the brain depending on the activity and problem involved.

Our brain is not a static organ. There is electrical activity between neurons in the brain that is constantly integrating new information coming from the external world. This natural process of react, adjust, learn, and remember has been passed on by evolution from one generation of humans to the next, and on to all of us in the 21st century.

"The study of memory might also affect pedagogy by suggesting new methods of teaching, based upon how the brain stores knowledge" (Larry Squire and Eric Kandel, *Memory: from Mind to Molecules*). Memories allow the brain to make predictions, if this, then that, the essence of intelligence.

Studies into memory have helped us understand how learning and memory operate within the brain. Memory can be learning that happens in four different stages:

1. Encoding information
2. Recalling: retrieving information not currently in our conscious awareness
3. Recognizing: identifying previously learned information
4. Relearning: we have deeper learning when learning information a second time

"Even though we have stored many things, we can only remember a few at a time and can only do so in a sequence of associations. Most of the information is sitting there idly waiting for the appropriate cues to invoke it. One set of neurons becomes active, which leads to another set of neurons becoming active, and so on. All memories are stored in the synaptic connections between neurons. This auto-associative memory system is one that can recall complete patterns when given only partial or distorted inputs." (Jeff Hawkins and Sandra Blakeslee, *On Intelligence*; 73)

> **A computer's memory is exact; the memory in the brain's neocortex is different. We do not perfectly remember what we see, hear, or feel. Not because human memory is error - prone, but because the brain remembers only important rela-**

tionships, independent of exact DETAILS. It recalls general concepts or what Jeff Hawkins calls invariant representations. We learn and understand not because of specific memories, but because of what invariant representations correspond to in general ways.

Approaches to learning should be geared for delivering information to a student's memory, using general concepts. Teach, coach and make suggestions using words and examples that remind students of what they already know, as the starting point for sharing new information. When possible, use metaphors, stories, and acts of play to share new information.

It can help our understanding of the nature of learning to realize that we are not driving the car, **our non-exact memory is**. We are not playing a sport; **our non-exact memory is**. For the most part, our memory is doing all these things with non-conscious predictions, free of exact details, supporting meaningful learning for future use. This type of behavior is also referred to as "motor" memory. It involves the structure in the base of the brain called the cerebellum. The heavily practiced motor movements such as playing a musical instrument, riding a bicycle, handwriting, typing, texting, dancing, and so on, are firmly encoded through repetition in motor memory. Happily, once these movements are learned, we do not have to think about the details of performing them.

For example, when we enter our car, the exact distance our hand will be from where we put the key into the ignition is different every time. A pianist and guitarist know how to position their fingers for the proper chords, automatically. The same applies to the auditory memory of a song in which we recall the important relationships in the song, not the actual notes. **Our memory has a general concept, and we unconsciously fill in the rest**.

Memories are formed and stored in the form of general concepts (not details) that the brain uses to make conscious and non-conscious predictions with. A combination of current input and stored general concepts will make detailed predictions, if this, then that is what is happing as we interact with changing environments. The methods used in approaches to learning should leverage this insight.

Memories are reconstructed; they are not recalled exactly as they occurred.

Two general categories of memory: "declarative memory" which refers to information, and "procedural memory" which refers to motor skills.

Within these two general categories, there is "intrinsic" motivation to learn in which the source of new information is mostly from self-discovery. This kind of learning is based mostly on a student's prior knowledge, and being self-motivated by a love of learning.

There is also "extrinsic motivation" to learn in which the source of information is external, delivered by an educator, parent, employer, coach, book, instructor, or the media and environments. During extrinsic motivation students are often seeking external rewards for their learning, the greatest seem to be high grades.

There are a variety of subjects anyone can learn, and there is declarative and procedural motivation, from intrinsic and extrinsic sources; but keep in mind that the gateway to learning and the starting point for most everything else we do is our brain and the emotional connection to the process of change. Learning is an emotional, chemical event.

Memory—A Primer. From the Dana Brain Organization, *May 9, 2013*

The brain's ability to change with experience gives us our memory, a faculty nearly as essential to mental life as breathing is to physical survival. Memory is the bedrock of our sense of self and the world, and our ability to manage the moment-to-moment demands of existence.

Thanks to a century of research, we know a good deal about its operation: what happens in the brain when we store facts, experiences, and skills in memory; what happens when we recall them. We can map in some detail the structures, circuits, and molecular processes underlying memory.

> **Memory systems:** Thoughts, perceptions, emotions, and actions are generated by groups of neurons firing together; memories are the patterns this activity leaves behind. There are two systems to preserve such neural networks:
>
> **Explicit** or **declarative** memory is what we can recall consciously and describe verbally. It encompasses two types of memory, sources of both mostly external. *Episodic* memory refers to the data of specific experiences: the sights, smells, sounds, and feelings of a day at the beach, for example. *The semantic* memory comprises what we have learned about the world, such as the meanings of words and objects, public facts like the name of the U.S. President during World War I, and information of personal relevance like the appearance of a friend's face.
>
> **Implicit** or **non-declarative** memory, gained mostly from self-discovery, is for things we do without thinking and typically do not put into words. The most familiar manifestation is a *procedural* memory, used to perform automatic actions like riding a bicycle or tying shoelaces. Habits and conditioned reflexes also rely on implicit memory.
>
> **The stages of memory:** Much of the time, we need to store information only briefly—while dialing a phone number or reading a paragraph from beginning to end. The brain accomplishes this with short-term memory, which holds data for seconds to several minutes.
>
> Scientists use the term *working memory* to describe the faculty for thinking about things while holding them in short-term memory, or switching between tasks that use

short-term memory (i.e. multitasking). Taking notes—listening to a speaker while writing down what he or she said 10 seconds earlier—uses working memory, as does talking on the phone while reading e-mails.

If we need to retain data for longer, it is stored in *long-term memory*. This is a multistage process that unfolds over time. Initially, data from an experience or bits of factual information are simply encoded—the brain selects and connects a few key elements for storage. As days, months, or years pass, the memory is *consolidated*, i.e. integrated with other experiences and facts into the framework of things one knows. The process of consolidation establishes memories more firmly, making them less subject to misremembering or forgetting.

While distinctions between short-term and long-term memory have traditionally been applied to the explicit memory, similar processes are at work in implicit memory as well. The formation of explicit long-term memories engages the hippocampus and surrounding structures in the medial temporal lobe, which connect input from different parts of the brain that respond to an experience or register a fact.

As consolidation occurs, the memory linkages become independent of the hippocampus: they are stored in the circuits of the brain that originally processed the information.

Implicit long-term memories are registered via direct modification of areas of the brain (cerebellum, motor cortex, striatum) that regulate movement, without the involvement of the hippocampus or any analogous mediating region.

Molecules of memory: Both short-term and long-term memory modify the synapses through which neurons connect with one another. This process typically strengthens existing synapses, but may also establish new ones.

Short-term memory temporarily alters neurotransmission. The release of neurotransmitters into the synapse starts a biochemical cascade of enzymes and energy-transferring compounds that increases the flow of charged particles into participating neurons, making them more excitable. It lasts only for the brief period that the released molecules remain active.

Long-term memory changes the structure of the synapse. A key process in this remodeling is *long-term potentiation* (LTP), which increases the number of and activity in certain receptors of neurons joined in the synapse, making them respond more strongly to neurotransmitters. LTP initially rearranges and reshapes bonds between proteins in the synapse. Further consolidation of memory modifies gene expression in the participating neurons to synthesize new proteins that strengthen memory-encoding synapses in a more enduring way. Sleep appears to play a key role in memory consolidation, by promoting the underlying chemical processes.

Remembering and forgetting: To use stored memories, they must be retrieved. Among the factors that determine how readily we can recall a memory are the conditions surrounding its initial storage: we are more likely to remember a situation, face, or fact later if we paid close attention to it at the time particularly if we had the intention of committing it to memory and if we had a strong motivation to remember it.

Emotion has a powerful effect on memory. Because strongly affecting experiences activate the amygdala, they register in memory quickly and deeply; some put emotional memory in a class by itself and are readily recalled. All too readily, sometimes: fearful or traumatic memories can be intrusive, as in phobias and post-traumatic stress disorder (PTSD).

Preservation of a memory is affected by how well it is integrated with other facts and experiences already stored in the brain. The more links with your general body of knowledge, the more reliably it will be recalled. Long-established memories may be retrieved more easily than new ones for this reason. On the other hand, memories tend to decay over time, particularly if they are rarely recalled.

Forgetting can also reflect active processes; new memories may interfere with old ones, and we may be motivated by the desire to suppress painful emotions or events. Selective erasure of memory may be as important as recall for effective mental functioning, and researchers are mapping out the molecular processes that make it possible.

RECAP

The nature of learning is like museums where you are guided by your own curiosity to have interesting interactions. Interactions cause active access to new insights and development.

Meaningful learning comes out of us, not into us.

Meaningful learning, more a product of wide observation, than the result of direct focused thinking.

Many traditional approaches to learning have misled students by doing most of the thinking for them.

When discussing Learning, include the elements of memory, emotions, organization, context, environment, experiences, methods being used, and the brain. Learning is a group or team effort.

In a student-centered/brain-compatible environment for learning students will: question assumptions, seek reasons, be reflective, make connections, draw conclusions, learn from experience and observations. In these environments, students contrast, evaluate, and express personal perspectives.

In student-centered/brain-compatible environments students also use self-reflection; foster creativity; are comfortable because it is a safe learning environment.

YOUR NOTES ____
____
____

CHAPTER ELEVEN

Closing Thoughts

"It is time to become familiar with research on learning; it is a positive message for all of us." Renata Caine, Professor of Education at California State College

"Education is the progressive realization of our own ignorance." Will Durant

ENCEMENT

"If you have attended many commencement ceremonies in your life you have likely struggled to stay awake the entire time," writes Gary Adkisson (Publisher, The Sentinel). On the other hand, Dean James Ryan's Commencement speech to the class of 2016 Harvard Graduate School of Education went viral on YouTube™ and Facebook™ and rightfully so.

The following are some highlights of that speech. Recognizing the importance of asking good questions Dean Ryan presented five of his "truly essential" questions he believes we should ask ourselves regularly to be both successful and happy. **This speech came synced with the point in which *Learning with the Brain in Mind* was being prepared for print; I found that these questions relate to the nature of learning and have included them here to demonstrate those relationships.**

Ryan's first truly essential question is 'Wait, what?'

"You have likely heard this asked by your children when you say something that they feel they must process to understand–something crazy like, 'you cannot turn on the television until you complete your homework.' The **'Wait'** means you have their **attention** and they are asking you to slow down for clarification; the **'What'** question indicates inquiry, and **inquiry always precedes advocacy**. It is at the root of all understanding."

The second essential question is, 'I wonder why/if.'

This shows **curiosity and a willingness to improve on an idea or thought**. It is at the heart of all curiosity. As you have read earlier, curiosity is at the foundation of all learning experiences. You have been curious enough to read this book. Now it is time to address the challenge of evolving how you can make the environment of approaches to learning more meaningful. Ryan challenges graduate educators with, "I wonder why students are so bored in school, and I wonder if we can make the classes more engaging?"

The answer is an emphatic YES; we can make learning opportunities more engaging! The value of our natural human need to explore and be curious should not go unnoticed or unprotected. Looking for your own solutions supports learning more efficiently than when answers are provided by others.

Dr. Yazulla's insights and suggestions were invaluable to this book. As mentioned earlier, he points out that what is obvious to us now in 2017 is that all our behavior, including learning information or skills, inputs from the senses, outputs to the muscles, and the coordination of these activities take place in the brain. Hopefully, reading *Learning with the Brain in Mind* assists those without cognitive training in gaining more insights of the value of curiosity and the basic principles behind the brain's internal connection to learning that supports gaining a meaningful education.

Emotions: Can you envision a classroom where a teacher creates an atmosphere that is so emotionally safe that when children are asked how they arrive at the answer for the addition of 2 + 3 all the children feel safe to raise their hand and share their individual way of getting to 5 (i.e. "I use my fingers to count," "I memorized it," "I tap my pencil to the count of each number while counting out the progress as I go," etc.)? Or a coach encouraging training time be spent on a topic being attempted in a variety of different ways to find the most effective outcome? Or being asked by an instructor in a genuinely curious, non-judgmental manner what steps were taken to come to a conclusion without critique during or after the delivery, but instead a compliment and encouragement to continue to analyze the steps. As that old song says, "oh wouldn't it be loverly, loverly, loverly" to feel that accepted and respected in your learning environment.

Emotions are a huge component of our ability to experience meaningful learning. How students feel about the components of where, how, from whom, and what they are learning matters. When educators lead by making a student's emotional response to learning opportunities their first concern, meaningful learning can become a reality in S.A.F.E., PLAYFUL environments.

Dean Ryan suggests a third truly essential question; "Couldn't we at least ..."

He says that phrase indicates a willingness to reach a consensus. It is at the beginning of all progress. "Couldn't we all agree that we care about the welfare of students even if we disagree about strategy?" "It is important to understand an idea before you advocate for or against it." The brief discussions here on some past approaches to teaching remind us of how far research has come in recent years and point out to some that **there are new horizons to consider**.

What can we successfully build on from research on approaches to information delivery? Recognizing that the Learning Sciences are uncovering insights that should not be overlooked when it comes to learning.

Because of contemporary research, by now in 2017 we have a reliable window into the concepts involved and the value of student-centered/brain-compatible learning environments. Several fields of study have conducted research that has led us toward expanding the curriculum in Education Degrees to include introductions to the Learning Sciences and their effect on a learner's ability to obtain long-term learning.

These studies point to the impact of culture and customs, emotions, memory and recall, neurological influences which are components of the nature of learning that are often overlooked.

What have we learned from the past that we should now consider avoiding? One suggestion; avoid creating what are referred to as "grade or outcome teachers" who are limited in view of learning. They look for what is right or wrong, not for what is being learned. They are teaching and fixing to get something right.

Teaching-fixing to get it right environments tend to be about a student's present skill level. They sometimes spray information at us without making contact. Whereas, learn, develop, and grow approaches are more about where students are going, and the ultimate outcome can be thrilling.

Teaching-fixing environments often mute curiosity and dilute imagination with passive, non-invasive approaches to learning. On the other hand, efficient student-centered learn-and-develop environments are safe and active, as individuals learn more from ever-changing environments by using real-time interactions with basic information.

Dean Ryan's fourth truly essential question is, "How can I help?"

"Asking how you can help indicates a **level of concern and caring** that goes a long way toward consensus and understanding. It is at the base of all good relationships." In writing this book, I hope to have led readers in the direction of at least considering taking advantage of contemporary knowledge from the Learning Sciences. I wanted to challenge readers to step beyond the past and continue to evolve and adjust approaches to learning based on what research has and is uncovering regularly, which I refer to as a student-centered/brain-compatible approach to teaching.

Ryan warns about the "savior complex; the expert or hero who swoops in to save others. We shouldn't let the pitfalls of the savior complex extinguish one of the most humane instincts there is, the instinct to help others. But how we help matters as much as that we do help, and if you ask 'how can I help?' you are asking with humility for direction, recognizing that others are experts in their own lives and that they will likely help you as much as you help them."

I have advocated for many years that I learn as much, if not more, from my students than they from me. We, players and coaches, are involved in our learning experiences together so asking a student how you can help might be a worthwhile bonding experience to free the student from any possible earlier learning environment hangovers.

Dean Ryan's fifth truly essential question is, "What truly matters to me?"

Dean Ryan suggests this might be asked when considering New Year's resolutions and refocusing on what things are truly valuable and matter in life. It gets you to the heart of life. Reflecting on the realization that not everything is going to go as we would like or how we may expect, we should ask ourselves "Did you get what you wanted out of life, even so ..."

"The 'even so' evaluates our aspirations, hopes, and dreams through a lens of maturity that recognizes life must be measured by love and respect." Ryan concludes by saying the world will be a better place not when people feel successful, but when they feel beloved and respected. In my opinion, these two are part of the foundation to being free to release one's curious desire to continue to evolve, learn, develop, and grow.

When it comes to the topic of education, be it as teachers, coaches, business leadership, or parenting, by continuing your pursuit of adapting to contemporary student-centered/brain-compatible approaches you are respecting those learners in your charge and will in return gain their respect.

Learning with the Brain in Mind is about beginning a journey into having the ability to influence meaningful learning via information-delivery methods that are compatible with the nature of learning. The benefit of this approach being long-term retention by the receiver. Having basic knowledge about the ever-evolving progress of the nature of learning has been found to be essential for enhancing learning opportunities. The interdisciplinary issues within the sciences of learning address how people learn, adapt and develop within the context of their environment and cultural customs contexts.

"How do the environment and individual cultural differences influence learning and development?

How can we best use emerging insights to design educational experiences and environments that improve learning and social-emotional development?

How can we best assess learning and development, evaluate the effectiveness of instruction, and conduct and analyze research that extends our knowledge base of these processes?

APPROACHES TO LEARNING AND DEVELOPING

Courses have a finish point, whereas learning never ends!

20thCentury	21stCentury
Teach	Educate
Passive learning	Active learning
Teach-conquer	Guide-stimulate-enlighten
Memorization	Development of learning skills
Critical Observations and Judgemental Outcomes	Provide positive feedback, encourage self-development experimenting
Focus on a specific result	Focus on learning process
Follow teacher's directions	Seek choices. Free to have undesired outcomes to grow from

20thCentury	21stCentury
Achieve to someone else's view of get-it-right	Achieve to a personal potential
Point out failure	All outcomes are future references
Teacher has the answer (biased)	Stimulate students to develop answer (open)
Teacher is evaluator	Self evaluation
Teacher uses how-to directions	Student encouraged toward self- discovery and experimentation

With the ever-increasing cognitive demands and the accelerated pace of change in the modern world, understanding and improving learning and development have become both increasingly challenging and vital to thriving in globalized and information-rich societies.

We know people are influenced by environments such as their home, neighborhood, cultural community, and school or place of work, including the virtual environments provided by emerging technologies. These are the environments in which learners develop and function, and they affect what they learn, how they learn, and how they use what they learn. Understanding these processes can help us develop environments in which children and adults can learn more effectively and live more enriched lives." (RMC website) Hopefully, what has been presented in this book helps readers gain insights into this topic.

Approaches to learning should be geared for delivering information to a student's memory, using general concepts. Teach, coach, and make suggestions using words and examples that remind students of what they already know as the starting point for sharing new information. When possible, use metaphors, stories, and acts of play to share new information.

In the end, information should be provided to learners in a manner that takes into consideration that compatibility is the key.

> When compatible with the nature of learning, the words used are sensitive to negative feelings, shared in context and free from cultural misinterpretation.
>
> They are compatible with the environment that encourages the learner to ask questions and probe forward in the direction of their choice.
>
> They are compatible with an approach that allows learners to feel emotionally safe in a supportive environment that encourages evolving to a conclusion.
>
> Compatible to incorporate the benefits of a playful environment.

I am in complete agreement with the statement that "Inquiry always precedes advocacy." Readers of *Learning with the Brain in Mind* may have begun the inquiry pursuit. As an educator, my personal and ongoing journey has caused me to recognize that "curiosity and a willingness to proceed with a thought are at the root of all learning" is an accurate statement!

My own curiosity will always lead me in the direction of making deeper inquiries in my journey toward finding more efficient ways of guiding learners around me toward a more complete and productive learning result.

BEFORE CONCLUDING

Before concluding this discussion on *Learning with the Brain in Mind*, let's retake the questionnaire from the beginning of the book.

Based on your current views, please evaluate the following statements as "mostly" true, or "mostly" false. These are the same questions as those in the beginning of the book. This retake allows you to compare your responses before reading this book to those after completing the book. Remember, for this exercise to be most effective, please consider your response to be your current opinion, not a response aimed at passing a quiz!

1. Meaningful learning environments are social environments. T F

2. Having students learn details can be less useful than helping them develop approaches to learning. T F

3. We should try to fix unworkable outcomes when learning. T F

4. We learn best from whole concepts, patterns, and sequences, not details. T F

5. The non-conscious mind is not valuable when learning. T F

6. Long-term learning is encoded after a lesson, and often during sleep. T F

7. Expert models are best used as models to copy. T F

8. Praise can be punishment in learning environments. T F

9. A master of anything was first a master of learning. T F

10. Actions of parents, employers, coaches, and educators should make individuals who are learning feel capable. T F

11. Being taught what a perceived expert believes is correct is less useful than students devel-

oping their own view. T F

12. A structured approach to learning is more useful than a random approach. T F

13. Learning often makes unconscious shifts (aha! moment) from a belief to a fact. T F

14. When learning, both workable and unworkable outcomes have value; with workable outcomes more useful. T F

15. New learning is encoded as the safety of the environment, and the emotional state of students are simultaneously evaluated by experiences stored in the brain. T F

16. Memory is more an act of rebuilding than recalling. T F

17. We did not have to learn to play; we play to learn. T F

18. When learning and performing, outcome goals are more useful than having learning goals. T F

19. The task of an instructor is to make students less dependent on them. T F

20. To be wrong is part of the process of gaining understanding. T F

21. It is better to be right, than curious with an open mind. T F

22. Knowing information is not the same as being able to use that information. T F

23. The human brain is wired to protect us from danger; physical and emotional. T F

24. Our prior ways of thinking do not impact new learning. T F

25. It is important that students avoid doing the wrong thing. T F

26. Positive thoughts improve attention. T F

27. Trying to suppress doing the wrong thing often recreates it. T F

28. Emotions can support or suppress learning and change. T F

29. Experience influences our thoughts in the now, and can stop or support learning and change. T F

30. The brain has an unlimited capacity for encoding information, but a limited capacity for recalling information. T F

31. Training and studying in one place for a period of time is more useful than several shorter time-frames and frequent change of locations. T F

32. Introducing errors when learning, supports experiencing meaningful learning. T F

33. The brain stores through similarities, but retrieves by differences. T F

34. It benefits learning more when notes are handwritten rather than typed. T F

35. The more information is distilled, broken down, and analyzed by instructors and students; the more it supports learning. T F

36. Intelligence comes before the behavior. T F

37. Information in the brain travels between neurons at over 200 mph. T F

38. Intelligence and understanding start as a memory system that feeds predictions into the sensory system. If this, then that. T F

39. Prediction, not behavior, is the basis of intelligence. T F

40. The brain self-learns; computers are programmed. A computer has to be perfect to work; the brain is flexible and tolerant of failures. T F

Answers:

1T, 2T, 3F, 4T, 5F, 6F, 7F, 8F, 9T, 10T, 11T, 12F, 13T,14F, 15T, 16T, 17T, 18F, 19T, 20T, 21F, 22T, 23T, 24F, 25F, 26T, 27T, 28T, 29T, 30T, 31F, 32T, 33T, 34T, 35F, 36F, 37T, 38T, 39T, 40T.

My Journey

Over time my approach and attitudes toward providing instruction to students and the process of experiencing meaningful learning have dramatically evolved. Initially, I used to do what other instructors suggested, a standard technique of teaching while stressing the basics. However, eventually, it became evident to me that with that approach some students did not reach their full potential or the level of competence that I believed they were capable of achieving. I felt something had to change.

During the first half of my career, I had the good fortune to receive some recognition and awards for my work as an instructor. Why then did I want to make changes when it appeared to some I was an effective teacher? Because studies showed that the approach to learning I and others were using would be described as "nature-of-learning poor." During those early years, my approach to teaching was subject content rich, which I would learn, is only one side of the story when it comes to learning. How information is delivered is the other side.

For years, I spent a great deal of time and resources gathering subject information from books, workshops, seminars, journals and from observing other teachers, as part of my efforts to grow as a teacher. However, I started to realize that I needed to be more informed about what supports or suppresses a student's ability to learn. I began to recognize that I was uninformed about how important emotions and the brain's connection to learning are.

> **I was clearly limited as a teacher because of my narrow understanding of learning and the positive and addictive quality of self-discovery. This insight caused me to investigate ways to change the strategies and substance of the information-delivery system I had been using for two decades, ones that I was realizing caused meaningful learning to slip away.**

One day, I asked myself what makes one approach to learning more efficient than others? Then I had an "aha" moment saying to myself, " how can any approach to learning expect to be effective without taking the nature of learning information and skills into consideration first."

Over the last 20 years, I attended numerous international seminars and workshops on the Neuroscience of Education both as a presenter and as a participant. At these meetings, I became informed about the many advances in understanding brain functions, how learning affects the brain and in turn, how both emotions and the on-going changes in the brain affect learning opportunities. Happily, I found that the research about the brain's connection to learning is more straightforward and understandable than I thought it would be. Today my aim is to provide emotionally safe student-centered/brain-compatible learning environments.

Student-centered/brain-compatible approaches can be applied to learning any topic. It is a process that opens curiosity and challenges the mind of students. I now see myself as less of a teacher and more as a guide, a facilitator, or a "learnest" who playfully supports an individual's ability to learn and perform up to his potential. A practitioner of science is known as a scientist, when learning or helping someone learn, to me you are a "learnest."

I, along with others, reasoned that the idea of providing a "playful learning environment" was a student-centered process and compatible with emotions and the brain's connection to the nature of learning.

Now, I often start a session with students by asking them to share what they already know about the topic. This approach gives me some insights into their current views of the subject and acts as a bit of pre-test. This process also lets students feel that they are helping me learn about them.

I now recognize that the time spent with me is not the lesson; it is a session. The lesson is what students take from what I have shared with them about learning the topic. What tools students take "from ' the session is based on their prior knowledge and how the information is shared. (FROM = **F**inding **R**elevant **O**bservations **M**eaningful)

At Harvard's Connecting Mind-Brain to Education Institute it was suggested that we should not see information as something we are giving to each other, but as information to think with, information to use, and information to think about.

I now use a process that guides individuals in the direction of inventing their own skills and personalizing their own ways of reaching workable outcomes. A student-centered/brain-compatible approach to learning is a process that is less about information and more about what students take from what is shared and can use in a variety of ways in dissimilar environments.

I now see the student as the real educator, providing information for both themselves and the instructor. Teachers who merely lecture or give out information are following past traditions that are often standing in the way of gaining an education. A teaching culture that does not help individuals find or invent their own knowledge base can actuality be a negative experience.

> I was slow to learn:
>
> Preconceived ideas about students can damage acts of learning, teaching, and performing.
>
> Being in charge is not collaborating.
>
> Trying to teach a subject often does not support a process of self-discovery learning.

Giving commands – does not provide guidelines and choices.

Trying to teach details using expert models is not incorporating a learning model.

Giving answers overlooks the value of guiding self-discovery and utilizing what the students already know.

A lesson is an opportunity to experiment, not a time to try to get it right.

Pointing out poor habits and failures does not support an emotionally safe, failure-free environment that enables recognition of valuable feedback for future reference.

Trying to fix unworkable outcomes does not help students change poor insights.

Trying to perfect a performance does not help students to learn to reach their own full potential.

Today, I understand when learning, developing and performing, that poor outcomes are natural inconsistencies. This was not my view for the first twenty years I was teaching; I saw unworkable outcomes as failure. Unworkable outcomes are natural components of every walk of life. **The aim in a student-centered/brain-compatible learning environment is for the student to become functional, without being misled by any idea of perfection.**

By becoming informed about the implications of the brain's connection to learning, we extend the boundaries of how the topic of learning has traditionally been looked at. For example, when it comes to meaningful learning; fiction influences facts, and facts influence fiction, and having information may not give us the answer.

I have learned that every action and thought we have is associated with some emotional component. Read that sentence again. To a large extent, what you do next depends on your emotional state of mind. Your decision, mode of action, and emotions are all organized in the BRAIN first. The ideal learning environment treats unworkable outcomes as opportunities, not failure and that view is reducing the negative emotional response that can accompany unworkable outcomes. Taking the brain's response into consideration in learning situations is the focal point of this book.

To Be Continued...

"We never know what the future holds, especially when it comes to what may be possible.

I know I will keep looking. Best of luck pursuing your goals."

Michael Hebron

About Michael Hebron

Michael was inducted into the PGA of America Hall of Fame in 2013. He was the 1991 PGA National Teacher of the Year and the 1990 National Horton Smith Award recipient for his contributions to PGA of Americas education programs.

Michael Hebron has received over 25 awards. He has been invited to India, Australia, Canada, Finland, Chile, Japan, Costa Rica, England, Ireland, France, Italy, Wales, Scotland, Switzerland, Czech Republic, Spain, Denmark, and as a private and public speaker to organizations, businesses, Yale, MIT and many other colleges, universities and school districts throughout the United States.

He has authored five books and over one hundred articles, has been a guest on several TV shows including the Charlie Rose Show and NBC's Today Show, has taken numerous classes at Harvard's Graduate School of Education Connecting the Mind-Brain to Education Institute and attended several conferences on Teaching with the Brain in Mind. Michael has been researching neuroscience and the brain's connection to learning with the help of award-winning educators and scientists for over two decades. What happens after a provider of information shares information with a receiver is his concern.

He provides continuing education scholarships at three high schools in New York (the High School of Public Service in Brooklyn; Holy Cross High School in Queens; and Smithtown High School on Long Island, NY).

If he had a choice, Michael would like to be known for how he delivers information to students rather than who the students he has spent time with are. Michael's preference is to pass on the value of playful, student-centered/brain compatible approaches to learning, which supports learning anything, not exclusively golf.

Michael Hebron holds membership in numerous societies including:

Learning and the Brain Society (LBS)

American Society for Training and Development (ASTD)

American College of Sports Medicine (ACSM)

National Center for Science Education (NCSE)

Association for Scholastic Curriculum Development (ASCD)

Phi Delta Kappa International Education Association (PDKIEA)

High School Coaches Association (HSCA)

Professional Golf Association of America (PGA)

United States Golf Association (USGA)

michael@michaelhebron.com http://www.michaelhebron.com

Acknowledgements

This book would not have been possible without the many forms of assistance I received. Not a day passes without my appreciation for a significant number of resources that have directly and indirectly influenced what I have compiled here. That this book bears only one author's name is unfair. I thank everyone who has helped me gain new insights into the nature of learning that I did not have during the first twenty years I was teaching. For the most part, I am sharing what I have learned from others about the brain's connection to learning.

A special thanks to the staff at Harvard University's Connecting the Mind-Brain to Education and its director Kurt Fischer. I also want to thank Eric Jensen, who along with his wife, organized several seminars about the brain's connection to learning that I attended. I see this book as a progress report on how my insights about learning and teaching have changed over time.

The content of this book was influenced by suggestions and questions from Professor Stephen Yazulla (Stony Brook University), Professor Terry Doyle (Ferris State University), Fran Pirozzolo (National Academy of Science on Techniques to Enhance Human Performance), Dr. Robert Bjork Director, UCLA Learning and Forgetting Lab. Elizabeth Bjork (UCLA), Andrea McLoughlin, Ph.D. (Associate Professor, Long Island University, Garth McShea, Karl Morris, Austin Curtis, Steve Orr, Nicholas Renna, Elayne Rattien, Richard Cohen, Susan Berdoy-Meyers, and every individual I have had the opportunity to help learn more efficiently. Thanks to Nannette Poillon McCoy and Dr. Stephen Yazulla for their ongoing assistance in copy editing and preparing the book for publication.

In the mid-1960s Gene Borek, a man I highly respect suggested that I write down my impressions of what I was reading, studying, and experiencing. Gene suggested that I write on a scheduled basis—daily, weekly, or monthly; the choice was mine, as long as I wrote. Gene felt that if I had a personal record of my thoughts and impressions, I would be able to look back and learn from my journey of change and development. Because of Gene Borek, who passed on in 2011, I have been making notes on a daily basis for over four decades. Thank you, Gene Borek, you are missed.

Everyone in modern science who has been researching the brain's connection to learning also has a big thank-you from me. Because of these men and women we now have a modern model that can enhance approaches to learning without having to know every aspect of how brains operate.

Many organizations and university research centers have revealed that there are some common assumptions about learning that are incorrect. These organizations include:

The Office of Education Research and Improvement of the US Department of Education.

Harvard University Graduate School of Education's Connecting the Mind-Brain and Education Institute, which recently committed to help improve learning experiences for everyone through research into the brain's connection to learning.

Vanderbilt University Peabody College of Education National Center for Quality Teaching and Learning, which is building new pathways for innovation, research, and reform in education.

University of Washington's Institute for Learning and Brain Sciences which is currently researching the brain mechanisms that underlie the windows of opportunity for learning.

Columbia University Teacher's College, which has the vision to have educational opportunities unlocking the wonders of human potential.

UCLA's Learning and Forgetting Lab, which has made great strides in uncovering insights in how we learn that are different from how we think we learn.

Bibliography

Ackerman, D. (2014). *The Human Age: The World Shaped By Us*. New York: WW Norton.

Al-Chalabi, A., Delamont, R.S. Turner, M.R. (2008). *The Brain: A Beginner's Guide*. One world Publications.

Ambrose, S.A., Bridges, M.W., DiPietro, M., Lovett, M.C., & Norman, M.K. (2010). *How Learning Works: 7, research- principles for smart teaching*. New York: John Wiley and Sons.

Angulo A.J. (2012). *Empire and Education*. New York: Palgrave MacMillan.

Arendt, H. (1981). *Life of the Mind*. New York: Houghton Mifflin Harcourt.

Arum, R. & Roska, J. (2011). *Academically Adrift: Limited Learning on College Campuses*. University of Chicago Press.

Ashton, K. (2015). *How To Fly A Horse: The Secret History of Creation, Invention, and Discovery*. Doubleday.

Augustine, Confessions; free e-book, www.gutenberg.org/files/3296/3296-h/3296-h.htm

Bailey, T.R., Hughes, K.L. & Moore, D.T. (2003). *Working Knowledge, Work-Based Learning and Education Reform*. Routledge.

Benassi, V.A., Overson, C.E. & Hakala, C.M. (Eds.). (2014). *Applying Science of Learning in Education*, Infusing psychological science into the curriculum. Retrieved from the Society for the Teaching of Psychology website: http://teachpsych.org/ebooks/asle2014/index.php

Bransford, J.D., Brown, A.L. & Cocking, R.R. (2000). *How people learn: Brain, mind, experience, and school*. Washington, DC: National Academy Press.

Bransford, J. (2007). *Understanding The Brain: The Birth of a Learning Science. Organization for Economic Cooperation* [OECD].

Brooks, D. (2012). *The Social Animal*. Random House Trade Paperbacks; Reprint edition.

Brown, Brené. (2012). *Daring Greatly*. Gotham.

Brown, P.C., Roediger III, H.L. & McDaniel, M. (2014). *Make It Stick.* Belknap Press.

Bruner, J. (1960). *The Process of Education.* Cambridge MA: Harvard University Press.

Caine, R. & Caine, G. (1994). *Making Connections: Teaching and the Human Brain.* Dale Seymour Publications.

Caine, R. & Caine, G. (1997). *Education on the Edge of Possibility.* Association for Supervision and Curriculum Development.

Carey, B. (2015). *How We Learn.* Random House Publishing.

Coyle, D. (2009). *The Talent Code.* Bantam Books.

Csikszentmihalyi, M. (2008) *Flow.* Harper Perennial Modern Classics.

Davidson, R. J. & Begley, S. (2012). *The Emotional Life of Your Brain.* Plume; Reprint edition

Dirksen, J. (2011). *How People Learn.* New Riders.

Doyle, T. (2008). *Helping Students Learn in a Learner Centered Environment.* Stylus Publishing

Doyle, T. (2011). *Learner Centered Teaching.* Stylus Publishing.

Doyle, T. (2013). *The New Science of Learning.* Stylus Publishing.

Drapeau, P. (2014). *Sparking Student Creativity.* Association for Supervision & Curriculum Development.

Dweck, C. S. (2007). *Mindset*, The New Psychology of Success. Ballantine Books; Reprint edition.

Emerson, T. & Stewart, M. (2011). *Learning & Developing Book.* ASTD.

Eagleman, D. (2012). *Incognito, Secret Lives of the Brain.* Vintage; Reprint edition

Esslinger, H. (2014). *Keep it Simple.* Arnoldsche Verlagsanstalt.

Feinstein, S. (Ed.) (2007). *Learning and the Brain, a Comprehensive Guide for Educators, Parents, and Teachers.* Rowman & Littlefield Education.

Feldenkrais, M. (1981). *The Elusive Obvious.* Meta Publications.

Ferrier, D (1876). *The Functions of the Brain*, https://openlibrary.org/books/OL24777733M/

Gatto, J.T. (2002) *Dumbing Us Down.* New Society Publishers; 2nd edition.

Gazzaniga, M.S. (2009). *Human: The Science Behind What Makes Your Brain Unique.* Harper Perennial; Reprint edition.

Ginott, H. (2003). *Between Parent and Child.* Harmony; Rev Upd edition.

Gleiser, M. (2014) *The Island of Knowledge: The Limits of Science and the Search for Meaning.* Basic Books.

Green E. (2014) *Building A+ Better Teacher*, W.W. Norton, and Co.

Harari, Y. N. (2015) *Sapiens, A Brief History of Humankind.* Harper.

Hart, A. D. (1996). *Habits of the Mind.* Thomas Nelson.

Hart, L. A. (1983). *Human Brain and Human Learning.* The village of Oak Creek, AZ: Longman Publishing Group.

Hawkins, J. & Blakeslee, S. (2005). *On Intelligence.* St. Martin's Griffin.

Hebron, M. (2009). *Play Golf to Learn Golf.* Smithtown, NY: Learning Golf, Inc.

Immordino-Yang , Mary Helen (2015) *Emotions, Learning, and the Brain: Exploring the Educational Implications of Affective Neuroscience, Volume 0 of Norton Series on the social neuroscience of education* W. W. Norton, Incorporated

James, W. (1899). *Talks to teachers on psychology: And to students on some of life's ideals.* New York: Henry Holt and Company.

Jensen, E. (2008). *Brain-Based Learning: The New Science of Teaching and Training.* Corwin.

Johnson, P. H. (2004). *Choice Words: How Our Language Affects Children's Learning.* Stenhouse Publishers.

Johnson, P. H. (2012). *Opening Minds, Using Language to Change Lives.* Stenhouse Publishers

Kahneman, D. (2011). *Thinking, Fast and Slow*. New York: Farrar, Straus, and Giroux.

Kaplan, M. & Kaplan, E. (2009). *Bozo Sapiens: Why to Err is Human*. Bloomsbury Press.

Kapp, Karl M. (2012). *Gamification of Learning and Instruction*. Pfieffer.

Kaufeldt, M. (2009). *Begin With The Brain*. Corwin; Second Edition.

Kawashima, R. (2005). *Train Your Brain*. Kumon Publishing.

Klemm, W. R. (1972). *Science: The Brain and Our Future*. Pegasus Books.

Kline, P. (1995). *The Everyday Genius*. Great River Books.

Knowles, M.S. (1975). Self -Directed Learning, a Guide for Learners and Teachers. Cambridge Adult Education.

Lemov, D. (2010). *Teach Like a Champion*. San Francisco: Jossey-Bass.

Levitin, J. (2014). *The Organized Mind*. Dutton.

Marcus, G. & Freeman, J. (2014). *The Future of the Brain*. Princeton University Press.

Mazur, E., (1997). *Peer Instruction, A User's Manual*. Prentis Hall.

Medina, J. (2008). *Brain Rules: 12 Principles for Surviving and Thriving at Work, Home, and School*. Seattle: Pear Press.

Mercer, N. (2000). *Words and Minds*. Routledge.

Mercer, N. (2008). *Exploring Talk in School*. SAGE Publications Ltd.

Moustakas, C. (1988). *The Authentic Teacher*. Irvington Press.

Nichols, M. (2006). *Comprehension Through Conversation*. Heinemann.

Painter, F.V.N. (1898). *The History of Education*. D. Appleton and Co.

Pascal, B. (1995). *Pensees*. Penguin Classics; Reissue edition.

Petty, G. (2014). *Teaching Today*. Oxford University Press; 4 edition.

Pearce, J. C. (1992). *The Magical Child*. Penguin Publishing Group.

Pollack, J. (2014). *Shortcut: How Analogies Reveal Connections, Spark Innovation, and Sell Our Greatest Ideas*. Gotham.

Race, P. (2010). *Making Learning Happen*. SAGE Publications Ltd; Second Edition.

Rath, T. (2006). *Vital Friends: The People You Can't Afford to Live Without*. Gallup Press.

Rechtschaffen, D. (2014). *The Way of Mindful Education*. W. W. Norton & Company.

Ringleb, A.H. & Rock, D. (2013). *The Handbook of NeuroLeadership*. Create Space Independent Publishing Platform.

Rock, D. (2009). *Your Brain at Work*. Harper Business.

Ryle, G. (2000). *Concept of Mind*. University of Chicago Press.

Sawyer, K. (Ed.) (2006). *Cambridge Handbook of the Learning Sciences*. Cambridge University Press.

Schacter, D. (2002). *The Seven Sins of Memory: How the Mind Forgets and Remembers*. Mariner Books.

Schmidt, E. & Rosenberg, J. (2014). *How Google Works*. Grand Central Publishing.

Sinek, S. (2011). *Start With Why*. Portfolio. Reprint Edition.

Smith, F. (1988). *Insult to Intelligence*. Heinemann.

Sousa, D. (2012). *Brainwork, The Neuroscience Behind How We Lead Others*. Triple Nickel Press.

Squire, L.R. & Kandel, E.R. (2000). *Memory: from Mind to Molecules*. W. H. Freeman.

Swaab, D.F. (2014). *We Are Our Brains*. Spiegel & Grau.

Tomassello, M. (2014). *A Natural History of Human Thinking*. Cambridge, MA: Harvard University Press.

Wagner, T. (2010). *The Global Achievement Gap*. Basic Books; First Trade Paper Edition.

Wagner, T. (2012). *Creating Innovators*. Scribner.

Waitzkin, J. (2008). *The Art of Learning*. Free Press.

Washburn, K. D. (2010). *The Architecture of Learning: Designing Instruction for the Learning Brain*. Clerestory Press.

Watts, A. (1989). *The Book: On the Taboo Against Knowing Who You Are*. Vintage Books.

Whitehead, A.N. (1955). *The Aims of Education*. Mentor (New American Library); Underlining edition.

Whitmore, J. (2002). *Coaching for Performance*. Nicholas Brealey Publishing; 3rd Edition.

Willingham, D.T. (2009). *Why Students Don't like School*. San Francisco: Jossey-Bass.

Wisk, M.S. ed. (1997). *Teaching for Understanding: Linking Research with Practice*. Jossey-Bass.

Zadina, J. N. (2014). *Multiple Pathways to the Student's Brain*. Jossey-Bass.

Zull, J. (2002). *The Art of Changing the Brain*. Stylus Publishing.

Recommended books on the topic of language:

> *Comprehension Through Conversation* by Maria Nichols,
> *Opening Minds, Using Language to Change Lives* by Peter H. Johnson
> *Words and Minds, and Exploring Talk in School*, both by Neil Mercer
> *Choice Words* by Peter H. Johnston

RESOURCES ONLINE

The Society for Neuroscience (sfn.org) is the largest organization of neuroscientists in the world. The Society homepage has a section called "Public Outreach" in the top bar.

Brainfacts.org (http://www.sfn.org/public-outreach/brainfacts-dot-org) is:

> A public information initiative of The Kavli Foundation, the Gatsby Charitable Foundation, and the Society for Neuroscience.
>
> BrainFacts.org contains a series of informative booklets on the structure, function, ethical issues, current research on the nervous system, and more. It is a resource for the general public, policymakers, educators, and students of all ages. BrainFacts.org is

dedicated to sharing knowledge about the wonders of the brain and mind, engaging the public in dialogue about brain research, and dispelling common "neuromyths."

Also available are links to Education Programs (http://www.sfn.org/public-outreach/education-programs) with resources for educators, and activities for educational organizations to participate in "Brain Awareness Campaign" (http://www.sfn.org/public-outreach/brain-awareness-week).

"Eight Conditions for Motivated Learning" by Kathleen Cushman in *Phi Delta Kappan*, May 2014 (Vol. 95, #8, p. 18-22), www.kappanmagazine.com;
Cushman can be reached at kathleencushman@mac.com
Neuroscience Online: an electronic textbook for the neurosciences. University of Texas Medical School. http://neuroscience.uth.tmc.edu/s4/chapter07.html

Squire, LR et al. The medial temporal lobe. *Annual Rev Neurosci*. (2004) 27: 279-306.

Different Facets of Memory. BrainFacts.org. Society for Neuroscience. 2012.
http://www.brainfacts.org/sensing-thinking-behaving/learning-and-memory/articles/2012/different-facets-of-memory/

The Brain from Top to Bottom: Canadian Institute of Health Research; Institute of Neurosciences, Mental Health, and Addiction.
http://thebrain.mcgill.ca/flash/a/a_07/a_07_p/a_07_p_tra/a_07_p_tra.html

Squire, LR, *Learning and memory—The Dana guide*. Dana Foundation website. http://www.dana.org/news/brainhealth/detail.aspx?id=10020

Hawkins, R. *Synaptic Plasticity and Learning. Basic and Translational Neuroscience*:
30th Annual Postgraduate Course, Columbia University. 2008. http://www.cumc.columbia.edu/dept/cme/neuroscience/neuro/topics/synaptic-plasticity-and-learning/

http://www.dana.org/News/Details.aspx?id=43548#sthash.4B2PEqty.dpuf
Link to Mary Helen Immordino-Yang on TED Talks www.youtube.com/watch?v=RViuTH-BIOq8 Link to Kurt Fischer
http://www.uknow.gse.harvard.edu/learning/learning002b.ht

Pittsburgh LearnLab: http://www.learnlab.org/
http://www.learnlab.org/research/wiki/index.php/Instructional_Principles_and_Hypoth eses

Johns Hopkins Science of Learning Institute: http://scienceoflearning.jhu.edu/

Institute for Educational Sciences: http://ies.ed.gov/
http://ies.ed.gov/funding/ncer_rfas/casl.asp

The International Society of the Learning Sciences:
http://www.isls.org/index.html?CFID=75710752&CFTOKEN=69962141

Center for Integrative Research on Cognition, Learning, and Education (CIRCLE), Washington University: http://circle.wustl.edu/Pages/Home.aspx

Human-Computer Interaction Institute Carnegie Mellon University:
http://www.hcii.cmu.edu/

Get Sports IQ.com TrainUgly.com

CPSIA information can be obtained
at www.ICGtesting.com
Printed in the USA
BVHW092312181218
535925BV00004B/389/P